Advances in Anatomy
Embryology and Cell Biology

Vol. 145

Editors

F. Beck, Melbourne   D. Brown, Charlestown
B. Christ, Freiburg   W. Kriz, Heidelberg
E. Marani, Leiden   R. Putz, München
Y. Sano, Kyoto   T. H. Schiebler, Würzburg
K. Zilles, Düsseldorf

Springer

Berlin
Heidelberg
New York
Barcelona
Budapest
Hong Kong
London
Milan
Paris
Singapore
Tokyo

A. Kriete

Form and Function of Mammalian Lung: Analysis by Scientific Computing

With 54 Figures

 Springer

Dr. ANDRES KRIETE
Image Processing Laboratory
Institute of Anatomy and Cell Biology
Justus-Liebig-University
D-35385 Giessen
Germany

ISBN-13:978-3-540-64494-1   e-ISBN-13:978-3-642-72220-2
DOI: 10.1007/978-3-642-72220-2

Library of Congress-Cataloging-in-Publication-Data
Kriete, Andres. Form and function of mammalian lung: analysis by scientific
computing / A. Kriete. p. cm. – (Advances in anatomy, embryology, and
cellbiology, Vol. 145)
Includes bibliographical references and index.
  ISBN-13:978-3-540-64494-1
1. Lungs-Computer simulation. I. Title. II. Series: Advances in anatomy,
embryology, and cellbiology; V. 145 QL801.E67 Vol. 145 [QP121] 571
s–dc21 573.2'519]

Production: PRO-EDIT GmbH, D-69126 Heidelberg
SPIN: 10672477   27/3136-5 4 3 2 1 0

Acknowledgements

This work has its origin in my Habilitationsschrift (professorial thesis) which was completed and submitted to the University Clinic Giessen in 1996.

First of all I would like to thank Prof. Dr. Dr. Hans-Rainer Duncker, Institut für Anatomie und Zellbiologie of the University Clinic Giessen, the main promotor of this work. His continuous interest and patience in counseling and his enthusiastic support both in the sciences and institutional infrastructure, combined with dedication and experience, have had great impact on the definition and realization of this thesis. His organizational, conceptual way of thinking was a breeding ground for developing my research.

In addition, my heartfelt thanks go to a number of persons who gave me support, ideas and encouragement. In particular, I would like to name Klaus-Peter Valerius, Institut für Anatomie, Giessen, for loaning me histological sections and casts and Prof. Dr. Dudeck, Institut für Medizinische Informatik, Giessen, who provided useful comments during the final compilation of my thesis. Besides the former co-workers at my image processing lab, Tim Schwebel, Hans Erbe and Michael Maier, I owe a debt of gratitude for computational support to Michael Marko and ArDean Leith from the New York State Department of Health at Albany and Wolfgang Krüger and Rüdiger Westermann from the Gesellschaft für Mathematik und Datenverarbeitung (GMD) in Bonn-Bad Augustin. The institutional support of Silicon Graphics in Frankfurt, the Royal Microscopical Society in Oxford, the microscopy division of Carl Zeiss in Jena, Vital Images and Imatec in München and the Department of Klinische und Administrative Datenverarbeitung at the University Clinic Giessen is also greatly appreciated.

Contents

1 Introduction

1.1
Overview

The precise knowledge of the three-dimensional (3-D) assembly of biological structures is still in its origin. As an example, a widely accepted concept and common belief of the structure of the airway network of lung is that of a regular, dichotomous branching pattern, also known as the trumpet model. This model, first introduced by Weibel in 1963, is often used in clinical and physiological applications. However, if this concept of dichotomy is used to model lung, a shape is obtained that is quite different from a real lung. As a matter of fact, many previous quantitative morphological and stereological investigations of lung did not concentrate on the spatial aspect of lung morphology but delivered data in a more statistical fashion.

Accordingly, the functional behavior predicted by such a model becomes questionable and indeed, the morphometrically predicted lung capacity exceeds the physiological required capacity by a factor of 1.3 up to a factor of 2. This problem has also been termed a paradox, as discussed by Weibel in 1983.

In the rare cases where descriptive models of the mammalian bronchial tree exist, monopodial in small mammals, dichotomous in larger ones, the understanding of the historical and/or functional reasons for size-related changes in the general design is not explainable. This investigation is trying to overcome this gap by computer modeling and functional simulation.

The oversimplification of the spatial structure of airways finds its continuation in the absence of descriptions of architecture of the respiratory units or acini. In addition to the limitations of stereological techniques, there is also an imaging problem. Acini cover large areas in histological sections, without exhibiting clear boundaries between adjacent units. Higher resolving lenses used in microscopy to image structural details such as alveoli limit the field of view, so that only small parts of acini are visible. Moreover, the limited penetration depth of light requires the use of hundreds of thin sections to cover a complete volume. For this the reconstruction of only half an acinus of a child's lung is reported by Boyden in 1980.

The past decade gave rise to the development of methods which focus on the investigation of the spatial structures. This complete different approach includes a 3-D imaging acquisition, 3-D computer graphical representations and 3-D measurement. The most important progress in microscopic imaging has been driven by the development of confocal imaging. With this system codevelopments in detector and laser technology, biochemistry and computer control were initiated. The use of these methods allows a new approach to investigate acinar structures as discussed in the first part of this work (Chaps. 2–5).

3-D analysis and computer graphics are then used to investigate the conductive part of lung. Recently, such computer based investigations have been used as a basis to simulate function, as discussed in the second part of this work (Chaps. 6–10). With the help of a complete morphological computer model of lung, simulations concerning the gas flow and gas uptake can be performed by computational physics. Such simulations critically rely on the boundary and initial parameters accessible. Lung research, however, is an area where such a procedure could be successful, since functional investigations such as dosimetry make these parameters available. Much advanced functional information in lung research was gained from all kind of dosimetric investigations, mostly conducted in humans, including variation of the breathing behavior, measuring gas uptake or analyzing the outcome of a gas bolus.

The fusion of form and function described here opens a new road for interpretation of the morphogenesis of lung. It is under discussion in how far generic physical mechanisms such as adhesion, tension, viscosity and convection may interact with genetic mechanisms and have an impact on morphogenesis and pattern formation. A set of computerized functional models of various species could certainly help to elucidate related questions in lung research.

The availability of a computer lung may also offer a number of practical, medical applications, in particular for the test of pollutant effects that underlie pollutant regulations such as highly reactive or toxic gases and particles. Their transportation and deposit into the lung can be studied by computer models. Simulations could be a surrogate for many animal experiments. Another area of application is in the development of inhaled pharmacological drugs for treatment of emphysema, asthma and fibrosis. To prevent underdoses or waste, aerosol drug deposition depending on particle size, density and ventilatory parameters could be studied in the future when computer models of the human lung become available.

1.2
Goals

For a thorough computer modeling of biological systems, the inherent structural hierarchies must be considered, since the macroscopic structures are built up from microscopic ones, and microscopic structures are based on macromolecular arrangements. No imaging techniques are known which transfer these hierarchies of biological structures directly into a numerical data format accessible by computers. There is some doubt whether such a technique will ever be developed because usually a large range of resolutions must be covered. This also holds for the organ of the mammalian lung, which is the subject of this investigation.

An initial set of hierarchical computer data representations is envisioned here to model the bronchial tree of a rat lung (see Fig. 1). Three levels of structural organization of the lung are taken into account: (a) the respiratory units (acini) at microscopic resolution, (b) the main stem macroscopic bronchi supplying for the lung lobes, and (c) the bronchial segments of the lung lobes. A bronchial segment is defined here as an individual airway tube between bifurcations. Each of the above-mentioned levels requires specific techniques for the imaging, analysis and modeling. The idea

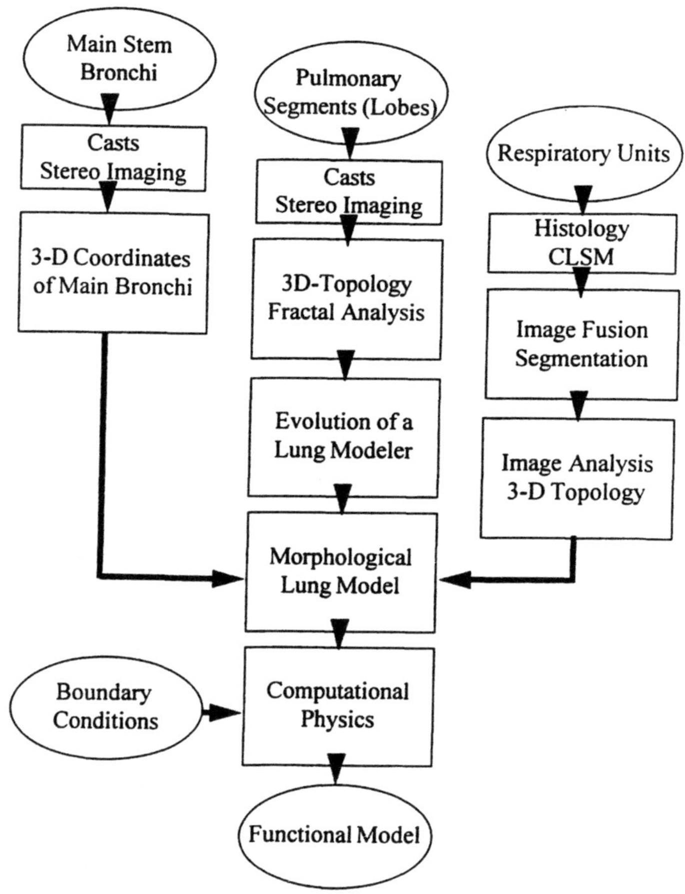

Fig. 1. Overview of the workflow required to explore a hierarchical, functional representation of a computer lung of small mammalien species. The microscopic investigation of respiratory units is subject of the first part of this work, the second part describes the investigation and modeling of the conductive part of the bronchial tree. Combining both structural hierarchies in a computer representations forms a model suitable for functional simulation by computational physics

behind these investigation is to model a realistic template of the bronchial tree. Stereo-imaging of casts and subsequent 3-D measurements are used to describe the main stem bronchi and the lobes.

For the study of respiratory units of lung, a confocal microscope is used as the primary imaging device for thick histological specimens. Since the confocal laser scanning microscope (CLSM) works with light, which is reflected, scattered or absorbed, the observable thickness is limited. Therefore, confocal microscopy was applied to serial sections of lung tissue. Subsequent image fusion of some hundred subimages sampled in precise and computer-controlled registration reveals the structural 3-D organization of these respiratory units at a size of about of $2\times2\times1$ mm with the alveoli clearly resolved. This method supports to the elucidation of the branching angles and architecture of acini, their space filling ability and their arrangement with neighboring acini.

Basic questions related to the study of acini include: (a) How can the inner structure of acini be accessed and visualized? (b) How can it be measured? and (c) How does a model of the acinus appear suitable for functional modeling? For the analysis of the conductive part of lung, a particular imaging problem must be solved: No technique is readily available to transfer the morphometric complexity of the pulmonary bronchial tree into a computer representation directly. It is, however, possible to access the bronchial tree by casts. Based on such casts, 3000–6000 segments must be measured at lungs of rat and mouse; frequently these have submillimeter lengths. Diameters range from 2 mm down to 0.3 mm. Even if 3-D laser scanners, such as whole body scanners, are improved for better resolution, many of the lung structures would overlap and hide other portions.

Therefore, casts of complete rat lung and individual lobes have been measured stereoscopically. The issue of this investigation is not to generate an average model in a statistical sense, but to represent a particular lung as well as possible. Therefore, main stem bronchi of a particular cast are directly transferred into a model as measured. The individual lobes which stem form the main bronchi are modeled by fractal graphics. As a consequence of over-simplified branching schemes idealized fractal models have been suggested, characterized by a fractal dimension. But, as demonstrated in this investigation, the individual handling of each lobe is important since they exhibit different branching patterns. In addition, the branching pattern can change within one lobe. For reliable functional predictions these peculiarities must be taken into account.

All three structural compartments, main stem bronchi, lobe segments and respiratory units, are finally combined into a hierarchical, 3-D model to represent a computer lung. This shows the significance and the potentials computers have in the modeling of biological structures, because they facilitate the integration of the structural hierarchies, multidimensional data and any variations over time.

With the help of a complete morphological computer model of lung, simulations concerning the gas flow and gas uptake can be performed by computational physics. Such simulations rely on the boundary and initial parameters reported by functional investigations. Within this novel functional model, flow and gas concentrations are known at any instance of the breathing cycle and in any bronchial segment. Indeed, all segments can be handled as individual, finite elements. Gas, or more precisely mass, is transported through each segment by convection or molecular diffusion or it is absorbed. The output of functional simulations can be displayed on the computer screen for better insight.

1.3
Scientific Image Computing

The computer-modeled representation of biological structures is progressing from the well-established analysis of structural information (3-D) towards the computing of spatio-temporal 4-D information (3-D+ time). Such opportunities are becoming of great importance in the life sciences, since variations of morphological patterns such as growth and dynamics are the main characteristic of living organisms. In addition to in vivo imaging techniques, 4-D information can also be gained from computer models.

Today, computer graphic visualization techniques are necessary to give insight into the avalanche of complex data and its variations. Although computer graphics and image processing have long been the province of computer scientists and engineers, it has recently reached across disciplinary boundaries to find new applications in interdisciplinary areas.

New imaging devices which profit from these development include the confocal microscope. Seldom has the introduction of a new instrument generated such excitement among biologists. This type of microscope allows optical tomography of thick specimens, has better resolution and contrast and gains impressive 3-D views of microstructures such as that of the lung tissue. The improvement in image quality of confocal over conventional microscopic imaging is a measurable quantity. The image quality criteria developed help to optimize the setup of the microscope. The confocal microscope plays a major role in what is characterized as the renaissance in light microscopy. In addition to optical technology, biochemistry, detector technology and computer technology are of equal importance for any progress in this area. CLSMs have become the chief alternative to electron microscopes in many biological research areas, since they offer the distinct advantage of observing structures nondestructively and in vivo where needed.

Visualization of sequential images is mostly carried out with a technique called volume rendering, which is not an easy task for lung tissue. Rendering of volumes is strongly driven by the imagination of the investigator of how the 3-D view of an object should look. But the realization of this imagination is often restricted by the capabilities of the soft- and hardware. For lung structures it must be made clear that in most cases the histological images are undifferentiated and they are compact – much as a sponge. The lung tissue could be rendered transparent, which of course would help only little due to the complexity and overlaps in structural composition. The way suggested here is in the computer differentiation or segmentation of structures such as bronchioli, ducts and alveoli and the rendition of these structures with different visualization attributes such as color and transparency. By these means, certain substructures can be tuned for 3-D viewing.

Due to the complexity of the mammalian lung in terms of number of structures, the underlying physical transport phenomena and the iterative mathematical treatment, functional simulations require a lot of computer power. Seen from the present state of the art in computer technology, workstations are capable of running a simulation within a few hours for small lungs such as rat and mouse for one breathing cycle. For more complex structures, such as the human lung, supercomputers or multiprocessor servers are needed. It is not the amount of memory required, but the enormous number of mathematical operations to be carried out which make such number-crunchers desirable.

Most of the programs discussed here have been newly developed in the C programming language, using a Silicon Graphics GL-library for all kinds of graphic displays. This includes renditions, measurements, the 3-D topological analysis, the various stages of the fractal models and the functional lung model, which includes computational physics. These programs were executed on SGI Indy/Indigo workstations or on a SGI Server Challenge L in cases where a higher numerical processing performance was required. Stereological measurements of casts were done with the Sterecon software developed at the New York State Department of Health at Albany and implemented on SGI computer systems. Preprocessing of the confocal images was achieved on specialized image processing hardware with programs developed from dedicated firmware tools (Kontron-Bildanalyse). Except for virtual reality applications, volume rendering was done with commercial software packages, in particular the Mipron software (Kontron) and Voxel View (Vital Images); the latter was interfaced to 3-D topological data.

2 Confocal Imaging of an Acinus

2.1
The Imaging Problem in Lung Research

The fundamental ventilatory unit in the design of the respiratory part of lung is the acinus. This term was first mentioned by Rindfleisch in 1878 (see Miller 1937). The term has been well fixed in the clinical literature since 1935 (Aschoff 1935). Various slightly different definitions about the proximal termination of this unit exist. It is, however, widely believed that all structures in the acinus participate, to a lesser or higher degree, in the gas exchange, whilst all structures proximal to the acinus are of air conducting nature only. Differences in this definition mainly address the gradual, transitional alveolarization in the bronchioles which are different in various species and the corresponding intra-individual developmental schemes (Valerius 1992).

In view of investigating complete units it seems to be necessary to define which structures of the bronchial part must be considered as belonging to the acinus, otherwise certain portions of the acinus might be missed and results might be difficult to be compared. Fortunately, the alveolarization of mouse and rat bronchioles starts very abruptly and segments called transitional bronchioli are not present in mouse and rat (Valerius 1992). Therefore, the end of the bronchiolus terminalis is defined here as the starting point of the acinus. For a given unit, this origin, together with the alveolar ducts, which are larger acinar pathways, and all other substructures attached to that compose the hexagonal acinus in total. Figure 2 depicts a cross-section through an acinus of a rat lung. Indicated are various bronchioli (B), alveolar ducts (AD) which end in alveolar sacs, areas mostly occupied by alveoli (A) and blood vessels (V). Numerous investigations based on cast models show that alveolar ducts are completely ensheathed with alveoli. Alveolar sacs are the blind-ending terminations of the alveolar ducts, which are also entirely surrounded by alveoli.

There is a long history of experimental difficulties in the quantitative investigation of the acinus of lung, because no technique was available so far to give highly resolved, complete insight into the structural complexity of an acinus. In order to do so, two imaging conditions have generally to be fulfilled for all kind of mammalian lungs: (a) the imaging technique should be powerful enough to penetrate a block of tissue containing a complete respiratory unit and, concomitantly (b), a resolution should be available to laterally and axially resolve the alveoli, which means 1/10 of the size of an alveoli or better.

For rats, the size of an acinus is in the range of 2000×2000×2000 µm. Depending on the degree of inflation, the diameter of alveoli is in the range of 40–100 µm for rats with a mean around 70 µm (Mercer et al. 1987), which requires a resolution of 7 µm or

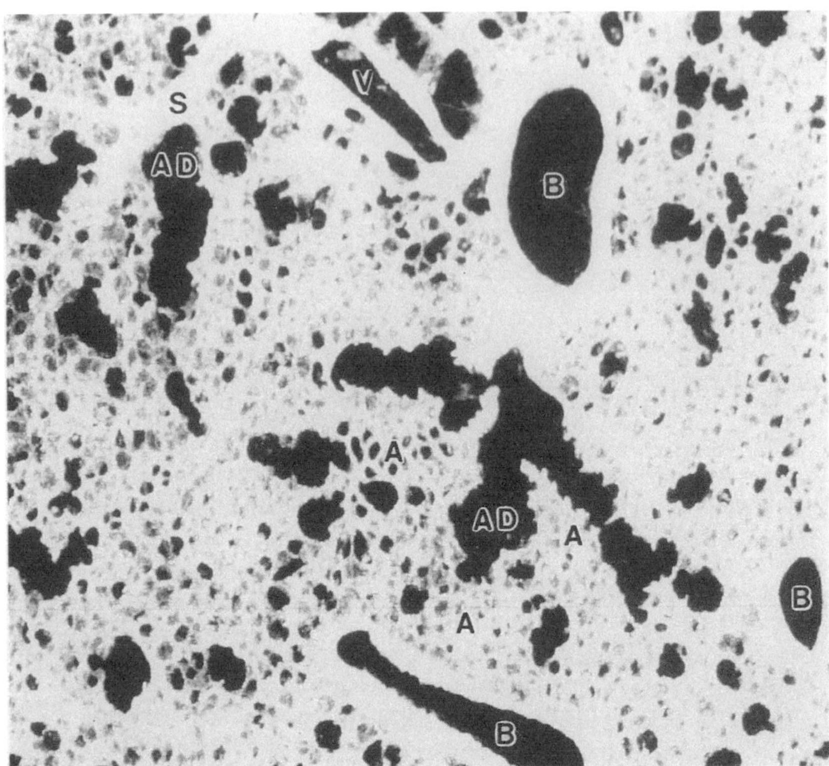

Fig. 2. Section of lung tissue imaged with a low magnifying lens (×5). Indicated are bronchioli (*B*), blood vessels (*V*), alveolar ducts (*AD*), a septum (*S*) and areas occupied by alveoli (*A*). Size of area displayed is 2×2 mm

better. At present, there are techniques such as X-ray imaging or magnetic resonance imaging which deliver a sufficient penetration power but lack the necessary resolution and contrast (Duke and Michette 1990). Investigations of acini are therefore mostly based on cast models for human lung (e.g., Haefeli-Bleuer and Weibel 1988) as well as for rat and mouse (e.g., Rodriguez et al. 1987; Valerius 1992). Unfortunately, these provide no insight into the internal structural organization but are restricted to a description of the outer form and a quantification of global parameters such as volume.

The only alternative is to mechanically cut a block of tissue into a set of serial sections and to investigate these with traditional light or electron microscopic techniques. Hereby the number of sections for a given block of tissue or the thickness of sections governs the lateral resolution. The disadvantages brought along with this procedure are obvious: It is not only a time-consuming approach, but what is required is a precise alignment and correction of the cut distorted histological sections. Furthermore, in view of the limited field of view of optical lenses a mosaic of images must be taken and fused to cover a sufficient field of view. The demands for the resolution required are still higher in this case. The problem is in the generation of "artificial" holes or pores between larger volumetric subtleties. Frequently, the volumes of adjacent ducts or even the air spaces between two adjacent acini are just separated by a thin parenchyma. If the fusion was not correct a slightly misalignment remains. Visual interpretation, but in particular automatic digital area or volume searching algorithms, might tend to generate these artificial pores. These passages may link structures together which might be misinterpreted as collateral ventilations. Because of these requirements, a complete 3-D computer-based investigation at a high level of resolution of an acinus of rat and mouse is not reported. However, one complete investigation of one half of a human acinus is given in the literature, based on a very time-consuming manually performed reconstruction of serial sections with a wax plate technique (Boyden 1971).

Confocal microscopy, in concert with digital image storage and visualization as an integral part of modern microscopy, can perhaps help to make an investigation based on serial sectioning technique more feasible and routinely practical. Every microscopic technique has its own limits, and a drawback which is shared by all microscopic systems is the fact that the observation is limited by specimen thickness and resolvable depth. Noninvasive optical sectioning overcomes these traditional limitations in optical microscopy and allows the acquisition of 3-D information in the form of sequential images. For practical applications, light absorption and scattering and the density of the tissue limit the total section thickness which can be observed.

It was suggested and demonstrated for single acinar pathways of the respiratory system that confocal imaging applied to thick sections and digital image processing might be promising (Oldmixon and Carlsson 1994). We describe a computer-guided image acquisition procedure in confocal microscopy, which allows a ninefold increase in the usually available field of view and an extended imaging in the axial direction by scanning aligned thick serial sections. This guarantees for the necessary axial and lateral resolution. Subsequently, a combination of serial sectioning and confocal microscopy is used to sample a large, highly resolved volume containing a complete respiratory unit.

For further interpretation, this complete volume of an acinus is visualized and quantified. Scientific visualization applied to 3-D microscopy gives rise to accurate and reliable concepts in computing and graphics to precisely interpret microstructures. Therefore, visualization and analysis are not mutually exclusive. This is quite obvious in the term "analytical graphics." In particular, any processing step to extract and identify volumetric substructures prior to quantitative analysis may be controlled by a 3-D reconstructed view. In addition, a number of visualization applications give evidence of integrated analytical tools to render specific volumetric subtleties. In order to fulfill such analytical tasks, visualization benefits from proven sets in digital image processing to enhance, filter, segment and code pictorial information, as required here to analyze the content of the acinar volume. As an inherent part of visualization the idea of topology is applied to measure the branching pattern and volumes of an acinus.

2.2
Material and Instrumentation

The specimens were taken from a half-adult lung of rat (*Rattus norvegicus*), donated by K.-P. Valerius, Institute of Anatomy, Giessen. At a mean weight of 237 g and a body length of 207 mm these animals are not very much different from the usually used

Sprague-Dawley rats. Under a pressure of 20 cmH$_2$O the inflated lung was removed from the animal and chemically fixed with Buoin's solution (containing picric acid) and dehydrated. Embedded in paraffin, mechanical sections of 70-μm thickness were cut with a Leica microtome equipped with a C-knife. These sections were stained with H&E; Hemalum was applied for 1 min and Eosin was applied for 30 s.

For imaging, an upright Zeiss confocal laser scan microscope was used (Zeiss LSM 410). The operation of this type of apparatus is explained in detail in Sect. 2.4. This system was equipped with two lasers, a green HeNe laser and a blue Argon laser. The system was controlled by a microcomputer (Intel 486). Images scanned are digitally accumulated in the image memory board (Matrox) and transferred to the hard disk of the system. We used a HeNe laser emitting at 535 nm. Fluorescence was detected with one photomultiplier above 545 nm from the lung parenchyma. Lung tissue, in its native form, gives rise to autofluorescence which stems from elastic fibers. The H&E staining is a histochemical staining procedure; the Eosin is an acid stain which links to the basic structures of the tissue. According to Romeis (1989), Eosin Y is identical with a fluorescein stain (Tetrabrom) and gives rise to additional fluorescence besides the red color visible at the cytoplasm and tissue. The cell nuclei, stained blue by the acid stain Hemalum, links to the phosphoric groups of the nuclei; consequently cell nuclei containing chromatin do not emit fluorescent light and remain dark in such preparations.

2.3
Prescanning and Definition of a Region of Interest

Microscopic investigation was started at a low magnification of a ×2.5 or ×5 lens. The ×5 lens (Zeiss Planapo 0.15) has a field of view of 4 mm^2. At this magnification prominent structures such as bronchii, arteries and veins and larger alveolar ducts were visible (Fig. 2). Using a confocal microscope at this stage had the advantage only that image quality was slightly better and more uniform than with a normal optical microscope equipped with a TV-camera for image capturing. However, the limited aperture of the ×5 lens was by far not high enough to give rise to any 'confocal' effect. Forty histological sections of 80 μm thickness each, located above and below a prominent cross-section given in Fig. 2, were scanned and stored digitally.

In order to get insight into the image sequence taken, the binarized and aligned images were contoured. Due to the high contrast, no special image-processing steps were necessary, but shift of the mean gray level in the images required an interactive control of the gray level threshold. Subsequently, the images were traced for structure boundaries, i.e., the demarcation between the lung parenchyma and the air spaces was marked. These closed polygons were subjected to a contour visualization program on a graphic workstation (see Chap. 4 for details).

Such a contour stack is given in Fig. 3. The overall size of the data volume in x, y and z is 2×2×3.2 mm. Such representations are difficult to interpret; much better is a surface representation, as discussed later (see also Fig. 28). However, based on a technique to find connections between contours (Chap. 4), a skeleton indicating the branching tree of this subvolume is also displayed. The approximate location and extension of individual respiratory units can be estimated for an investigation at higher resolution. There is a bronchiolus at the right, three terminating bronchioli (TB) and

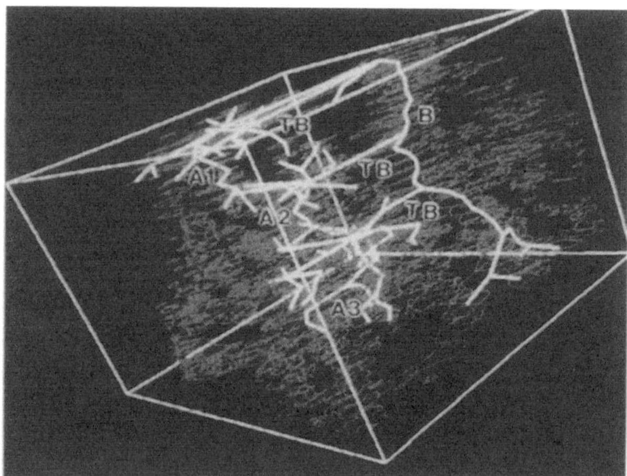

Fig. 3. Three-dimensional contour stack of the airways in a data cube extending 2×2×3.2 mm. With the help of connecting centerlines, the branching pattern is visualized. Three terminating bronchioli (*TB*) originate from a bronchiolus (*B*) at right and branch into acini (*A1–A3*) at left. The main ducts of these respiratory units are represented. This set, based on 40 serial sections, serves as an preview to identify and locate acini of interest

their starlike clusters of main alveolar ducts forming acini (A1-A3). Except for the analysis of major ducts, the low lateral and axial resolution and the digitization into a matrix of 512×512 pixels and the tracing process prevent any detailed investigation. Therefore, a technique is discussed below to increase resolution without sacrificing the field of view which will be applied to the respiratory unit marked A1.

2.4
Confocal Laser Scanning Microscopy

2.4.1
Basic Principles in Confocal Microscopy

In order to resolve thick histological sections, the opportunities of confocal microscopy were theoretically negotiated and some basic tests were performed. The fundamentals of confocal microscopy go back to 1959, when a new kind of microscopic apparatus was described (United States patent, M. Minsky 3013467, granted 1961). The confocal microscope, compared to a standard microscope, has enhanced lateral and axial resolution, improved contrast and, in particular, it can remove the out-of-focus blur of thick specimens. Thus, an optical sectioning capability is achieved. The final breakthrough of confocal microscopy in biomedicine and material sciences occurred 30 years later, when the commercial realization of the optical principle could also make effective use of powerful computers to digitally store, visualize and analyze image data. In addition, highly developed optoelectronic components such as laser light sources and sensitive detectors as well as the application of specific fluorescent dyes made a wide acceptance of this imaging technique possible. Details of the historical development of confocal microscopy are reviewed elsewhere (Inoue 1990; Carlsson and Lunddahl 1992; Cox 1993).

2.4.2
Types of Systems

Two different kinds of confocal instruments are available: the tandem scanning microscope and the laser scanning microscope. The *tandem scanning microscope* uses normal light chopped by a rotating Nipkow disk (Petran et al.1968). The light beams reflected by the specimen cross the rotating disk on its opposite side (thus tandem) and due to the symmetric pattern of holes out-of-focus blur is removed. This system allows real-time and real-color imaging, but is limited in sensitivity due to a fixed size of the pinholes. The more commonly applied CLSM features a variable pinhole and allows a flexible adjustment of magnification by modifying scan angles (Fig. 4). In addition, various imaging modes such as fluorescence, interference and polarization contrast are possible.

A schematic example of the principle of *confocal laser scanning* is given in Fig. 5. The laser light is focused by the microscopic lens (d) and illuminates a small spot in the specimen. Light reflected and rescattered or fluorescence is imaged by the objective and directed onto a photomultiplier (i) by a dichroic mirror (a). A pinhole (n) in front

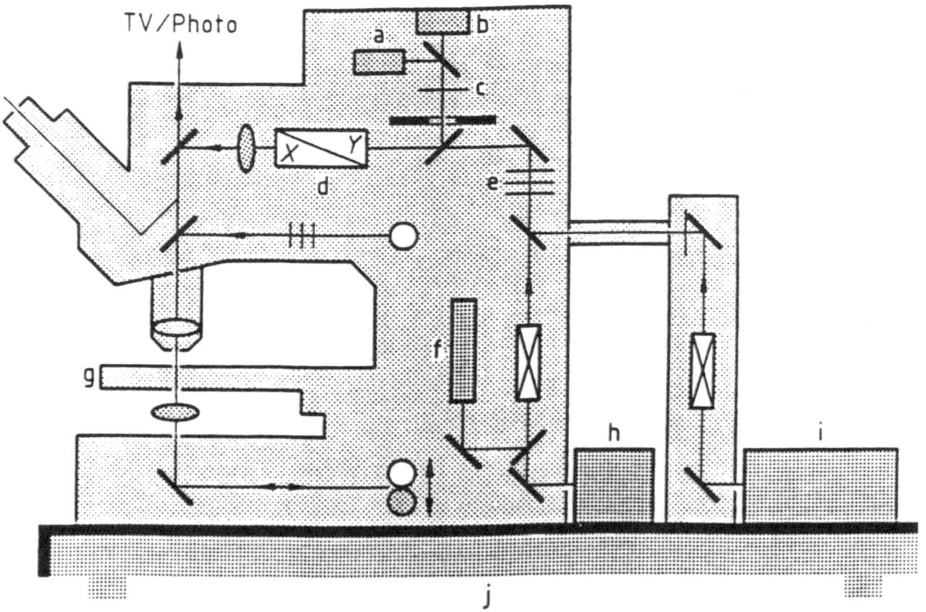

Fig. 4. Setup of a confocal microscope, based on a research microscope (Zeiss). *a,b,* Photomultiplier tubes; *c,* confocal variable pinhole; *d,* scanners; *e,* Barrier filter; *f,* internal HeNe laser (543/633 nm); *g,* stage; *h,* Ar laser (488/514 nm); *i,* optional UV laser (350–360 nm); *j,* antivibration table (with permission from Carl Zeiss)

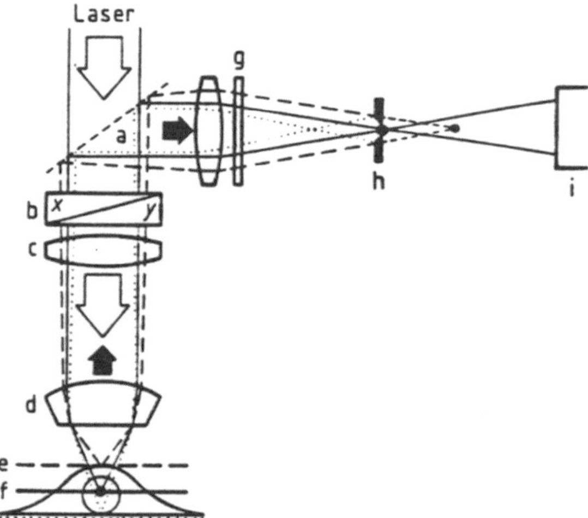

Fig. 5. Confocal imaging in laser scanning microscopy. The laser light is focussed by the lens (*d*) onto a specimen with certain thickness and scanned by mirrors (*b*). The light coming back from the specimen is captured by the lens, deflected by a dichroic mirror and focussed onto a photomultiplier (*i*). Light which stems from the specimen lying in focus passes a pinhole (*h*), but contributions from above (*e*) and below the specimen are filtered out (with permission from Carl Zeiss)

of the photomultiplier is positioned at the crossover of the light coming from the focused point. This plane corresponds with the intermediate image if seen from the Koehler illumination. Light emanating form above and below the focused point has its crossover behind and before the pinhole plane so that the pinhole acts as a spatial filter. Numerous papers elucidate confocal image generation (e.g., Sheppard and Choudhury 1977; Sheppard and WilsonT 1978; Brakenhoff et al. 1979).

Since only a small specimen volume is recorded, scanning must be performed by either moving the specimen (stage scanning) or by moving the beam of light across the stationary specimen (beam scanning). The latter is the solution realized in all commercial instruments. Theoretical considerations have shown that the two scanning forms give identical resolution. Blocking out-of-focus light at the detector pinhole generates an optical sectioning capability, which opens a way to scan a 3-D structure with a slight axial stepping of the stage (see Fig. 6).

2.4.3
Imaging Properties

The imaging characteristic in confocal microscopy is given by the specimen properties weighted by the applicable spatial confocal response function. An explanation based on wave optics by Lethonen et al. (1992) considers the point spread functions of the illuminating path and the detection path (resembling the photon detection possibility) to be multiplied, consequently resulting in a narrow point resolution.

One must differentiate a reflecting object and the corresponding coherent imaging properties from a fluorescent object viewed in incoherent imaging. In terms of

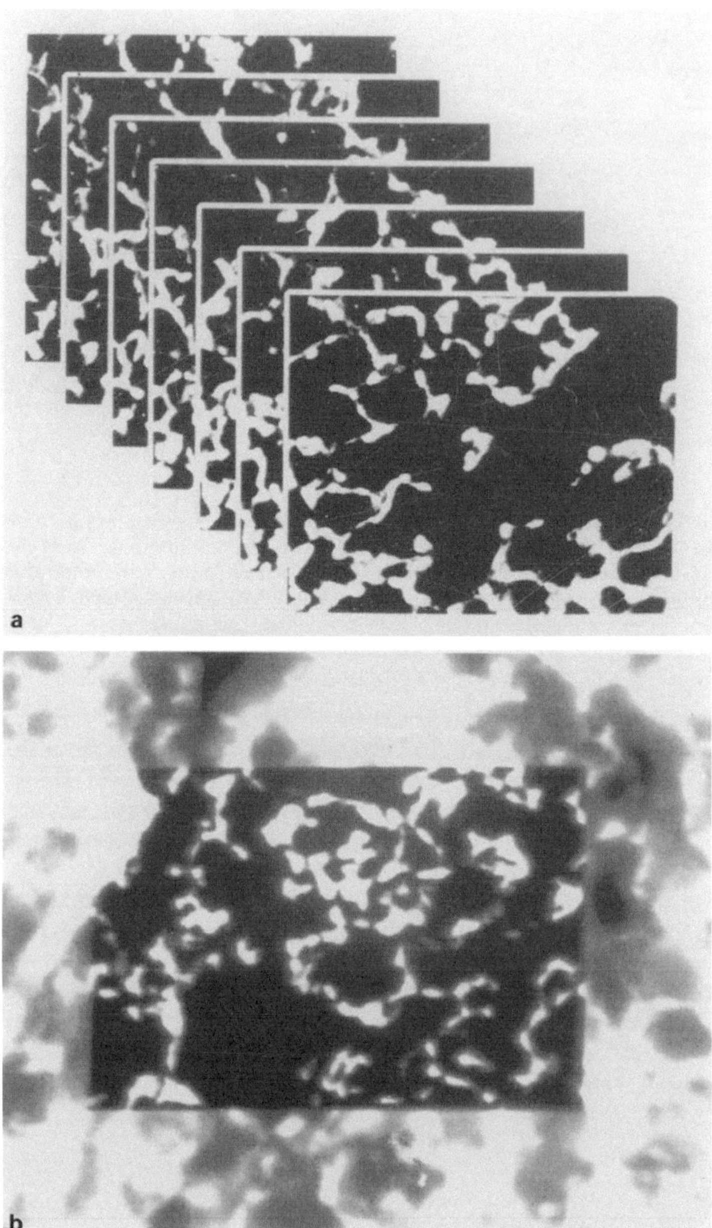

Fig. 6. a Multislice series generated optically with a confocal microscope. **b** Conventional versus confocal imaging at a region-of-interest of a lung tissue

resolution, the Rayleigh criterion determines the minimum resolvable distance between two points of equal brightness. For incoherent imaging using a lens with a certain numerical aperture NA at a wavelength l, the lateral resolution d(l) is (Eq. 1 Sheppard and Wilson 1978):

$$d(l)=0.46\ l/NA \tag{1}$$

Assuming a wavelength of 514 nm (green argon laser line) and a numerical aperture of 1.4, this equation gives a resolution of 157 nm. Compared to conventional imaging, where the resolution is given by the Abbe equation, the resolution in confocal microscopy is about 32% higher (Cox et al. 1982). Abbe's equation does not give a real physical limit of what is achievable in optical microscopic imaging, and in fact due to the point illumination and the pinhole detection the confocal microscope has a better resolution.

An approximation of the axial resolution d(a) is given by (Eq. 2):

$$d\ (a)=1.4\ l/NA \tag{2}$$

The values (514 nm and NA=1.4) give an d(a) of 367 nm.

One way to measure the axial response of a confocal microscope is to record light intensity from a thin layer with infinite lateral extension. In a conventional microscope no difference, and therefore no depth discrimination, is obtained. However, in confocal microscopy, there is a strong falloff if the object is out of focus. The free width half maximum (FWHM) of such a function resembles the optical section thickness (Wilson 1989). This size also depends on the diameter of the pinhole, and the optimal pinhole size (D) with a maximum deterioration of 10% in reflection is given by (Eq. 3; Wilson 1989):

$$D=0.95\ l\ M/NA \tag{3}$$

with magnification M, l and D given in micrometers. In fluorescence microscopy, depending on the ratio between excitation and emission wavelength, the size has to be reduced between 10% and 30%.

Another way to describe the imaging performance is the modulation transfer function (MTF); such measured and theoretically derived 3-D functions have been published (Sheppard and Gu 1992; Carlsson and Lundahl 1991). From an image processing point of view, the improvement of "confocal" image quality is a measurable quantity, as discussed in Appendix 1.

2.4.4
Confocal Instrumentation

Confocal microscopic equipment is available as an add-on to a standard microscope or in form of a more fully integrated and stable device, including the computer control equipment. Both upright or inverted types of microscopes are available. The inverted confocal microscope allows the study of larger specimens and easy handling of micro-manipulating or injecting devices. For both types of instruments beam scanning is obtained by rotating galvanometric mirrors. Axial control is realized by a stepping motor drive or a piezoelectric translator. Most instruments are equipped with several detectors to record multifluorescence events. Some also offer transmission detectors, and images obtained in different channels can be mixed electronically. Recently, variations of the confocal technique have been developed such as bilateral scanning in real-time with a scanning slit (Brakenhoff and Visscher 1992). Because of the fast scanning, parallel detectors such as CCD-camera can be used with such devices. Scanning slit systems offer a better signal-to-noise ratio but slightly less resolution than a CLSM.

Various types of lasers can be linked to the microscope (Gratton and van der Veen 1990), but up to now routinely only continuous wave gas lasers are used. This includes argon ion with excitation lines at 488 and 514 nm and helium-neon with a major line at 633 nm and a weaker line at 543 nm. The use of UV lasers requires achromatic lenses and specifically designed scanning mirrors. Lasers used in the UV-range are argon ion UV or argon ion tunable lasers, krypton ion and helium-cadmium Lasers. Depending on the emission lines of the fluorescent dyes, various filters are used in front of the photomultiplier detector (PMT), characterized by the spectral response. The total sensitivity of a system is given by the transfer efficiency of the various optical components (Sandison et al. 1995) and, in addition, the quantum efficiency of the detector and its associated electronics (Art 1990).

Simplified types of laser scanning microscopes, such as the fiber scanning optical microscope, have also been proposed (Sheppard and Gu 1992). A totally different, physical approach to reduce out-of-focus blur is based on a physical phenomena called double-photon excitation. Hereby, fluorescence is generated by simultaneous absorption of two photons of long wavelength at the site of a single molecule. The necessary energy emitted by pulse lasers is only present at the focal plane, thus a "confocal" effect is originated without any additional optical specialties (Webb 1990). Recently, sensitive dyes have been developed for three-photon absorption (Bhawalkar 1995; Hell 1996).

Recently, confocal imaging in transmission has also been studied. Improvement in contrast and resolution depending on the pinhole size of thin specimen could be documented in a confocal transmission arrangement at differential interference contrast (Cogswell and Sheppard 1992).

The diffraction pattern of the illuminating beam is elongated axially if compared with the lateral direction (see Eqs. 1, 2). This anisotropy is basically caused by the limited numerical aperture of the microscopic lens. An optimal diffraction pattern would be achieved by an unlimited, 360° aperture. This concept is also known as the 4-pi microscope. A prototype version uses two lenses centered around the specimen. The object in focus is illuminated from two sides and the fluorescence emanating from the object is detected through both lenses as well (Hell and Stelzer 1992).

2.4.5
Contrast Generation

Confocal imaging modalities include reflectance, rescattering and fluorescence (Cheng and Kriete 1995). The first results in confocal imaging have been obtained at unstained tissue in *reflectance* (Egger and Petran 1967). This is also the preferred imaging modality used in industrial inspection of semiconductors, often combined with methods such as optical beam induced conductivity (OBIC). In biomedical sciences, the observation of unstained tissues is very limited due to low reflectance but must be intensified by metal impregnation (e.g., golgi stain). This includes peroxidase-DAB labeling with nickel in neurobiology (Deitch et al. 1990) or silver stained nucleolar organizer region-associated proteins (AgNORs). Also gold immunolabeling (van der Pol 1989) has been used.

In *rescattering*, variations in the refractive index can give rise to a reasonable contrast. One particular example is in ophthalmology (Masters 1992) where structures in the transparent cornea such as cell nuclei, ceratocytes, neurons or the fibers of the lens can be visualized.

Fluorescence is the most commonly used imaging mode in confocal microscopy, for example, in cell biological applications (Shotton 1989). This includes the use of autofluorescence, specific dyes, in combination with antibodies and in situ hybridization. Comprehensive reviews of available fluorochromes are published elsewhere (Haugland 1989; Tsien and Waggoner 1989; Rigaut 1992). UV lasers broaden the spectrum of available fluorescence dyes, and even the autoflourescence present in reduced pyridine nucleotides [NAD(P)H] can be monitored (Masters et al. 1993).

New applications also rely on specifically developed fluorescent markers, such as biosensors or environmental markers. A fluorescent dye is chemically linked to a macromolecule, creating an analog to the natural molecule. It is specifically designed to measure chemical properties of the molecule, instead of tracking exact locations of macromolecules within the cell. Calcium-regulatory proteins are such a class of biosensors (Lansing and Taylor 1992). Also the mobility of molecules can be studied with a method termed fluorescence redistribution after photobleaching (FRAP). Hereby a small volume of the specimen is monitored after bleaching with a high light intensity. The recovery of fluorescence within the volume indicates the diffusion coefficient. In particular, the mobility of lipids and proteins at membranes have been studied (Lansing and Taylor 1989).

Optical sectioning devices, being noninvasive and having low radiation damage, are ideal for studying living specimens, organs and tissues. Confocal microscopes are such devices to observe structures in vivo at full 3-D microscopic resolution (Terasaki and Dailey 1995). Small organisms or organisms in an early stage of development can be observed, such as in embryogenesis. Concerning individual structures, morphological changes such as the movement of cell compartments, the growth of neurons or synaptic plasticity in the retina have also been studied (Kriete and Wagner 1993). Moreover, confocal arrangements allow an optical trapping, i.e., a user controlled movement of small spheres filled with biochemical solutions into and within cells.

An increasing field of new application of confocal microscopy is also in clinical imaging. This includes the observation of wound healing, flow processes in veins (Villringer and Dirnagl 1992) and ophthalmology (Masters 1992).

2.4.6
Application of Confocal Imaging to Thick Sections

Equation 2 suggests using high numerical lenses to achieve a confocal effect. If, however, a wide field of view is required as well, the necessary low magnifying lenses

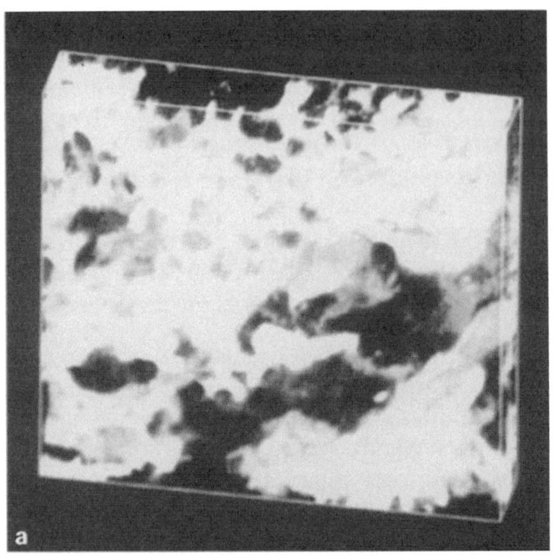

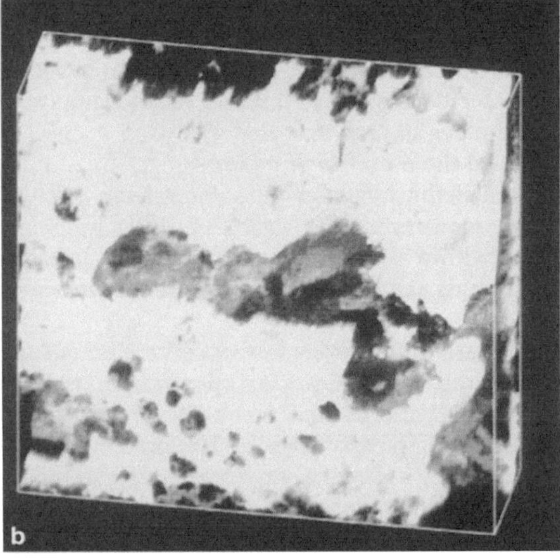

Fig. 7a,b. Visualization of a data set of lung scanned with a LSM, 15 (**a**) and 30 (**b**) optical sections are processed. Image size is 250×250 µm

have a performance not very much different from conventional imaging. As an example, at an H&E-stained lung specimen of 70 µm thickness, the mean intensity through focusing reveals a peak at the center of the specimen and no z-resolution is obtainable. A lens with a higher numerical aperture produces a plateau over a certain thickness of the specimen, provided the specimen has an isotropic structure. The reason for this effect is in the z-resolution of the lenses, or more precisely their half-width maximum of the diffraction pattern. If the z-resolution goes beyond the

18

section thickness of the specimen, a confocal effect called optical tomography starts to take place (see also Appendix 1).

The maximum thickness of a specimen which can be imaged depends on a number of factors, including the density of the specimen which affects the penetration of light into and reemission of light from the specimen and variation in diffraction index which produces scattering within the specimen and degrades image contrast. For lung structures, a maximum of 100 μm was found to be an appropriate limit without visible degradation, but this is also influenced by the digital sampling rate (or digital resolution) applied. In thicker specimens, image processing can, to a limited degree, correct these undesirable effects, but for a precise restoration complicated calculations must take place, since these effects are highly nonlinear, i.e., object dependent, and structures lying at the surface of the specimen effect the light path at the next deeper level and so on.

Figure 7 gives an example of a confocal data set. Based on 15 and 30 optical sections scanned with a ×20 magnifying lens a computer generated volumetric 3-D reconstruction is displayed (a method which is discussed in detail in Chap. 4). Bright structures correspond to the fluorescence of the lung parenchyma. Alveolar ducts and the package of alveoli are clearly visible. Due to the structural complexity, one cannot look inside the lung tissue, a problem that is similar to the drawbacks of corrosion cast models. However, comparing several data stacks exhibiting various numbers of sections of the same area helps to interpret the 3-D relationships.

Other undesired effects that occur in confocal microscopy include the fading of fluorescence, which was, however, not observed for this type of auto-fluorescing specimen. For specimens which suffer from fading, sophisticated, nonsequential z-sectioning can give more uniform results (Stevens 1990). Another correction important for quantification concerns the difference in diffraction index between the immersion oil and the specimen-embedding media. Any mismatch can effect the x/y to z relation of structures imaged throughout the specimen (Visser and Brakenhoff 1991).

2.5
A Framework for Scanning Large Volumes in Confocal 3-D Imaging

The previous sections have shown that the confocal microscope is used in the sense of an optical tomograph to acquire image sequences with precise registration in thick specimens. The axial resolution of such a device is improved with lenses of high numerical aperture, but unfortunately the lenses available today designs bring along a narrow field of view at the same time. Certain studies such as that of the lung parenchyma, however, require both a good axial resolution and a particular wide field of view. This conflict is resolved here by an image compositing technique which combines a sequence of thick sections; each of these sections is confocally resolved at several locations in an array like fashion and all the resulting subvolumes are digitally combined.

Since the image acquisition procedure for a wide-field, high resolved data volume is a laborious and time-consuming task, careful planning of the image acquisition is necessary. Scanning at various resolutions and aligning the image series from step to

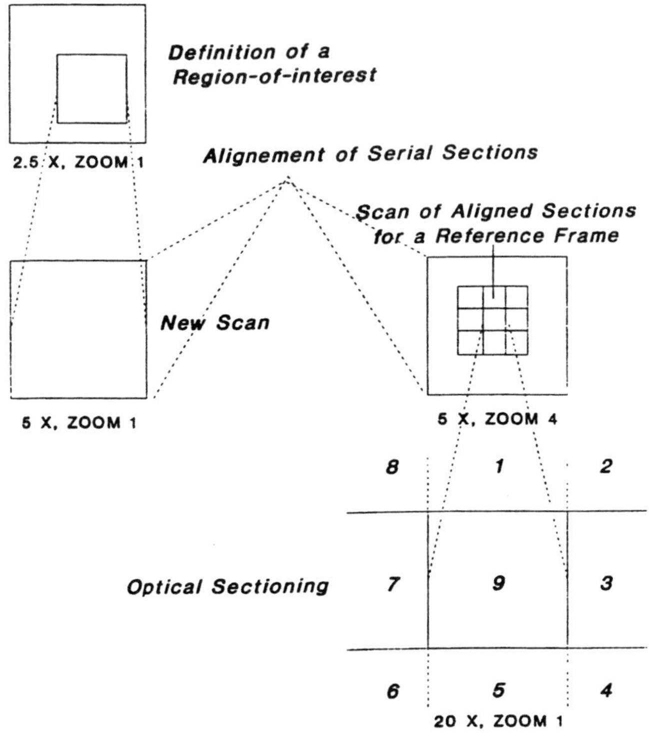

Fig. 8. Framework for a computer-controlled image acquisition of a large, highly resolved microscopical volume out of thick serial sections

step generates a framework which is necessary for orientation, reproducible scanning and control. This setup is discussed below and graphically displayed in Fig. 8.

Step 1: Definition of a Region Of Interest: An interesting acini, characterized by the branching pattern of ductus, was selected from a contour-based reconstruction (see Fig. 3, acinus A1). Once the region of interest was defined, series were scanned anew at the region of interest and centered to the structure. As mentioned before, the numerical aperture of this lens is not sufficient for a good axial resolution. These images were transferred to an image processing system (Kontron IMCO, see Appendix 2) and were aligned for linear shift and rotation.

Step 2: Alignment of Images: The main disadvantage and well-known problem in serial sectioning is the need for correcting lateral and rotational shifts, distortions due to mechanical cutting and influences of the staining processes summarized in the term alignment. Usually, just a best-to-fit procedure is used. This has two disadvantages: (1) the first section may be already severely distorted, so that all other sections following may be erroneously corrected and (2) the subjective decision biases structures selected for alignment and may bring along a trend during slice by slice alignment, so that the whole structure looks twisted when completed. If there is no way to work with artificial marks, such as laser fiducial marks (Bron and Grimellet 1992;

20

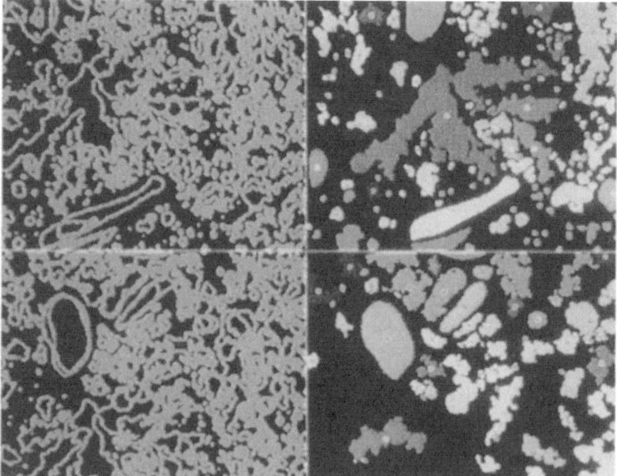

Fig. 9. Alignment of images. The binary images are identified (labels are applied) and the centers of gravity are determined. The corresponding centers determine the transformation matrix. The overlay of contours of the two sections documents the improvement of alignment

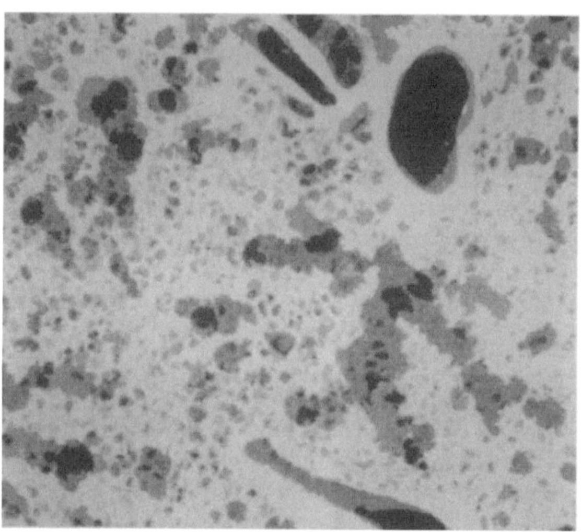

Fig. 10. The overlay of the LSM helps to pre-align sections and to re-align the specimen stage for a new scan with corrected reference images transferred back to the LSM

Kriete and Magdowski 1990) or if the cutting procedure does not allow images to be recorded from the block surface to serve as episcopic reference images to the diascopic microscopic images (Lamers et al. 1989), a more automatic procedure might improve the degree of precision. Using digital image processing helps to find landmarks objectively suitable for alignment and may also bring along the ability to quantify of how good the alignment actually is (Fig. 9).

For the set of lung images present, it was decided to use artificial landmarks in the form of centers of gravity. After segmentation of the images into binaries (see Chap. 3), the air filled spaces were handled as individual objects having a center of gravity. Finding the corresponding center of gravity in the next section made it possible to define transformation parameters of this particular point. The advantage of using such artificial marks over anatomical landmarks was that this procedure was less sensitive to structural changes from image to image and influences of variations in the segmentation process were minimized as well. Figure 10 gives an example of processed adjacent sections. The centers of gravity could be automatically extracted after image segmentation and labeling with random colors (left side). On the right side the result of the alignment before and after the processing is demonstrated by an overlay of contours.

At least three points, but mostly five points, had to be defined to make the transformation matrix robust, i.e., stable for new points coming in. The transformation matrix was then used for recalculation of the image matrix, and linear lateral shift, rotation and zoom were taken into account. Building the difference of both images gave a measure of the degree of alignment achieved. This, however, could not be generalized or compared with differences of other adjacent sections, since the difference depends on the content of the images. Transferred back to the confocal microscope, these corrected images serve for alignment of the histological sections during confocal scanning.

1Step 3: Realignment of Sections: In the next acquisition step, prealigned digital images were used as a reference image to scan a set of sections anew. For manual alignment, the real color display option of the LSM was used (Fig. 11). The actual image was displayed in the red channel and the reference image was kept in the green channel of the image memory of the image processing board of the LSM. Continuous update of the scanned image allows the interactive control of the alignment. Manual rotation and shift of the specimen led to a best fit of the two images, indicated by a yellow color of structures fitting.

Step 4: Setup of a Reference Frame: Next a mosaic of reference frames is defined at higher magnification using electronic zoom (×4). Variation of the scanning offset, i.e., shifting the scanning beam electronically away from the center in all eight possible directions at a lateral distance of 600 µm, gives nine reference images. At this zoom setting they construct a frame of 1800 µm side length (data hold for a Zeiss LSM 410). An example is given in Fig. 11a, showing nine images composed as X-windows images on the computer screen. What is obvious from this arrangement is that there is a slight overlap in y (vertical) direction compared to the x (horizontal) direction, caused by a difference in the electronic guidance of the scanning mirrors. Compared to Fig. 2, this image already reveals a slightly better resolution, mainly caused by a higher digital resolution.

1Step 5: Optical Sectioning of a Wide-Field Array: Switching from a ×5 to a ×20 magnifying lens (Plan Neofluar NA 0.5), the field of view at zoom factor 1 could be fitted exactly to each of the nine individual reference images. By manual lateral shift of the scanning stage the specimen area actually scanned could be fitted to a particular reference image of the mosaic. For axial setting of the scanning stage, focusing started on top of the specimen until a constant intensity was reached, which was usually about 5 µm from the very top. Subsequently optical sectioning could commence in this area and seven sections of 10 µm of thickness were scanned and stored digitally (an example

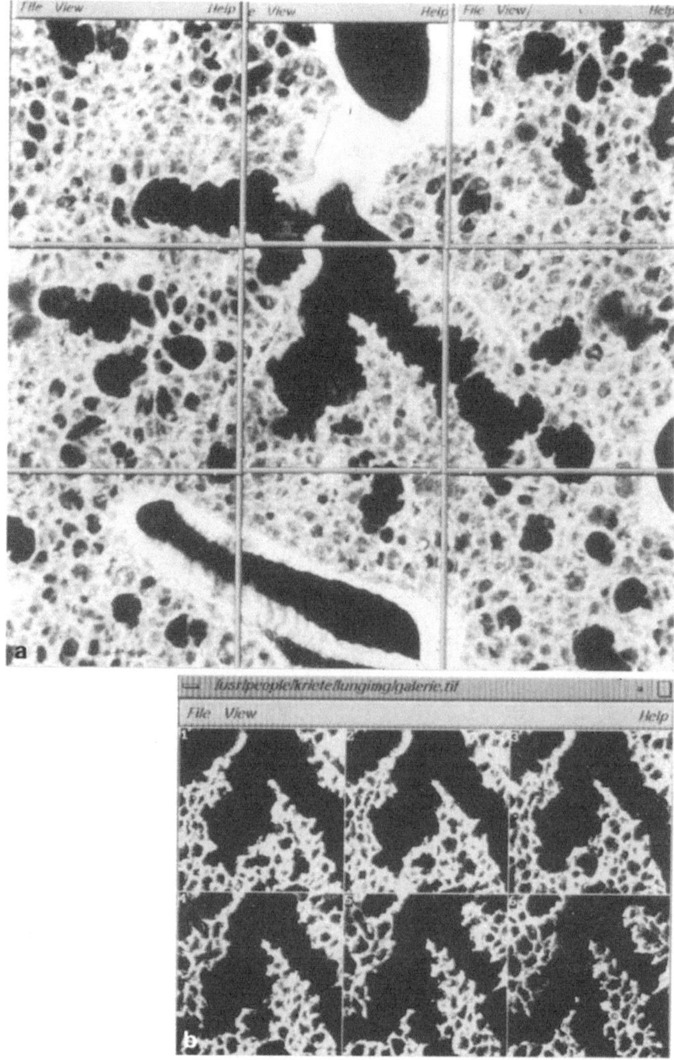

Fig. 11. a A mosaic of reference images composed at the workstation under X-windows. The image size of the individual reference images is 256×256 pixel. b Confocally resolved example of a subvolume used as a source for image fusion

is given in Fig. 11b). The compactness of the thick sections which appear when observed with traditional imaging is greatly deconvolved and simplifies the image processing discussed below. By this reproducible arrangement, for each of the nine reference areas nonoverlapping subvolumes could be acquired by optical sectioning.

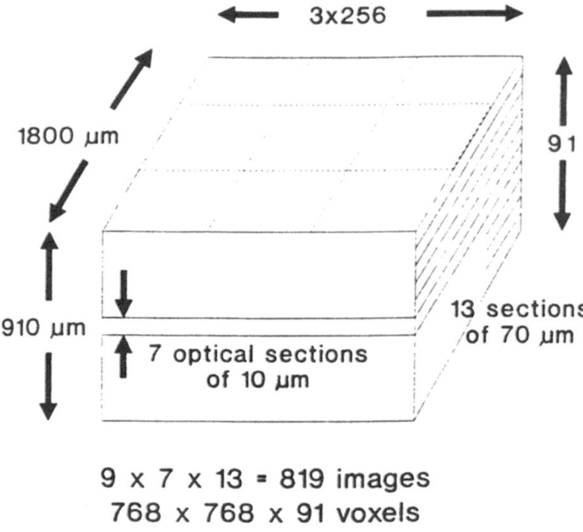

Fig. 12. Dimensions of the composed data set of an acinus

2.6
A 3-D Data Volume Representing a Complete Acinus

For the reconstruction of a particular acinus 13 thick sections of 70-μm thickness, each scanned at an axial step size of 10 μm, result in a total of 9×7×13=819 images of 256×256 pixels each. *Pixel* is an established abbreviation of the digital elements or picture elements that construct the image matrix. Subsequently, these images were fused at every level of optical sectioning. The digital final volume composed was of 768×768×91 elements being equivalent to 1800×1800×910 μm. Hereby the pixel can be handled as elements having a certain thickness to construct the volume. They are called volume elements or *voxels*. This particular volume has 53,673,984 voxels which are stored along with the intensity values of 8 bit or 256 gray levels in the computer. The volume encloses one complete acinus, but also exhibits structural details such as the alveoli. Figure 12 explains the dimensions present in this data set.

3 3-D Analysis of a Complete, Highly Resolved Respiratory Unit

3.1
Basics of 3-D Analysis

A special advantage of using computers is the ability to quantify image volumes. The classical method is stereology, based on the measurement of sections with special measurement grids. The application of this technique is, however, mostly limited to the measurement of randomly oriented and distributed structures.

Digital image processing, however, allows direct quantification of specific structures of any form and orientation in 3-D, but preprocessing steps to segment and identify structures are usually necessary. Subsequently, the measurement can be carried out and some of the procedures developed in digital image analysis are based on modified stereological algorithms.

The quantification can address size and volume of structures, quantities, or forms and relation between volumetric subtleties. To measure volumes all voxels which identify a certain structure are summed. It is particularly difficult to measure the surface of structures and most methods offer an estimation only based on stereological methods.

Object counting in volumes usually requires the definition of guard-boxes, which are moved throughout the volume. Objects falling into such boxes are evaluated following certain stereological rules in order to avoid any bias. Stereological considerations are also used to estimate spatial statistics, such as numerical densities or nearest-neighbor distances; such methods have been intensively used in 3-D cytometry (Rigaut et al. 1992). What is described in the following is an attempt to segment a lung data set, to measure some basic parameters and to label this data set as a preparation for 3-D visualization following (Chap. 4).

3.2
Image Preprocessing

If digital image processing is considered as a tool to quantify structures in three dimensions, undesirable phenomena due to the imperfections of the imaging process must be corrected first. In confocal microscopy, the attenuation of the laser beam at thick specimen, photobleaching or saturation of the fluorescence (Rigaut 1992; Brakenhoff 1992), autofluorescence and background noise, or the distortions caused by the embedding media (Visser and Brakenhoff 1994) are subjects to be taken into account. In very thick specimens, a loss of resolution may be observed. The correction of such phenomena is often limited by the underlying theoretical model available, which must describe bleaching phenomena, loss of resolution etc.. Moreover, the effects are highly object dependent.

A particular example in confocal microscopy is to correct undesired effects in ratio imaging, used for the purpose of cross-talk reduction, which occurs between two narrow emitting fluorophores recorded with dual detectors. Rationing, i.e., the compensation of two channels by digital division, can substantially reduce cross-talk effects (Wallen et al. 1992). Environmental markers indicating the PH-status or the degree of free calcium (Ca^{2+}) also require ratio imaging in order to compensate for structure densities or section thickness (Tsien and Bacskai 1995).

Due to the unisotropic lateral/axial resolution of the confocal data sets various ways to improve axial resolution are available; deconvolving methods use the point spread function of the optics, which must be measured first. The algorithms can be differentiated further by their ability to incorporate noise. The classical filter is the Wiener filter; new developments include the maximum likelihood estimation (Holmes and Liu 1992; see also Appendix 1) joint with a blind deconvolution. A different approach is the nearest-neighbor deconvolution developed for brightfield imaging (Agard et al. 1989), taking into account the blur present in the planes above and below the focus plane. More recently, this method has been contrasted with a no-neighbor deconvolution method (Wang et al. 1994). In addition to a direct improvement of the data, isotropy can also be obtained by interpolating the lateral dimension (Rigaut 1992). Transformation of data from a cubic format into other lattices have also been proposed, based on mathematical morphology (Meyer 1992).

To avoid some of the problems discussed above we have used 80-µm-thick sections which show no loss of resolution at the magnification used. The heterogeneous structure of the lung tissue keeps absorption effects at a minimum. With respect to autofluorescence of this tissue the bleaching effects are minimal. Any residual effects could be corrected with usual image preprocessing techniques, as discussed below.

Sets in digital image processing to enhance, filter, segment and identify structures are frequently applied to 3-D microscopic data sets. This can be done either section by section or by using the corresponding 3-D extension of digital filters such as the Gauss, Laplace, Sobel, and Median filter (Cheng et al. 1992; Van der Voort and Brakenhoff 1990; Mossberg et al. 1990). Most image filtering algorithms can be easily implemented in a higher-dimensional data space. Especially edge detection algorithms have been proven useful if extended to the third dimension.

Preprocessing pipelines have been proven useful in various applications, such as in electron microscopy (Kriete et al. 1984), but in particular to enable pleasing visualizations, such as in the investigation of complex chromatin arrangements (Montag et al. 1990). Different steps must be arranged in a way that they consider the increasing demands of the different 3-D visualization algorithms.

Here, the processing steps starts with a histogram equalization (see upper part of Fig. 13) This was necessary to harmonize the intensities within the parts of one composed section and in between the sections themselves. What was essentially done is a mapping of the occupied gray levels in the image to the maximum range of 0–255 in a way that an approximate uniform distribution was achieved. The resulting cumulative histogram distribution was linear. If applied to microscopic images taken under slightly different illuminations but basically from the same object, the gray level distribution can be mostly equalized and the images look very similar. Usually, the

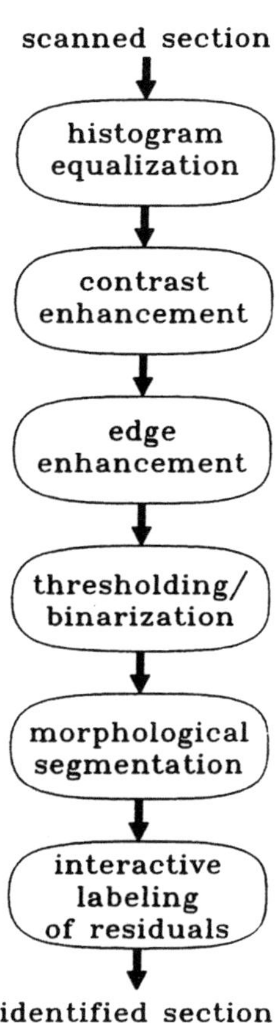

Fig. 13. Image processing pipeline, which consits of preprocessing, segmentation and identification modules. This pipeline has a strict top-down, section-by-section processing character. Structures are automatically extracted by morphological operations of the binary images

contrast of the image is enhanced as well. Next, the sharpness of the edges was enhanced, particularly at transitions between various phases of an image, similar to a highpass filter. Since the filtered image only indicates the transitions, it was weighted and combined with the original one. Matrix size was 3×3 and weighting was 3/16 of the filtered image and 13/16 of the original one.

3.3
Segmentation and Labeling

For the lung data set, the labeling of identified structures is a step to prepare measurements and structural attributes for 3-D visualizations. In particular, the image must be divided into lung parenchyma and background.

Segmentation often requires interaction by the user, since automatic algorithms almost rely on the intensities of the voxels only. Here, the threshold selected was interactively controlled, which is used as a demarcation between lung parenchyma and background. Due to the way of preprocessing, modification of this threshold was required in the range of 5–10 gray levels only.

Once binary images had been obtained, it was necessary to identify structures out of the images such as the outer boundary of the acinus, ductus, bronchi and alveoli (see also lower part of processing pipeline in Fig. 13). During the labeling the black and white pixels are replaced by user selected gray levels. To differentiate ductus and alveolar sacs completely from the alveoli, binary morphological operations using structuring elements for erosion and dilatation were performed (Conan et al. 1990). This procedure, if performed several times, removed all structures lying beyond a given size. Here, the filter size and the number of operations were selected in a way that mostly all alveoli, many of them in a sponge such as open form, could be differentiated from the ducts. By this procedure, about 80% of all structural details were identified. Since the exact boundary of the acinus is difficult to define, in particular in the outer areas some of the structures remain questionable and must be identified interactively.

3.4
An Automated Segmentation Procedure

The overwhelming structural detail of the data sets make the above described labeling method very time-consuming, since all the procedures must be applied anew for the section following until the whole data set is processed. What was tested instead was a still more automated procedure (Fig. 14). First, one representative image out of the middle of the data stack was selected as a starting point. At this image, an interactive guided labeling of structures as described above was performed. Next, this identified image was used as an marker or reference for the following one. What was actually done was that the image following was randomly identified by the computer, which means all structures or objects present and separated from others are detected and labeled by a randomly selected color. If combined with one selected label of the reference image, this label was transferred to those objects which overlap by at least one pixel of the area with the reference object. Due to changes from section to section not all the structures were identified correctly, and some interactive correction was always necessary. For this particular data stack, the procedure was carried out three times up and down the complete data stack, in order to fill up all nonidentified areas. A typical result is the section given in Fig. 15.

A particular problem again occurred at the boundary of the acinus. Due to structural changes from section to section, artifacts from the cutting procedure and errors in the segmentation, an automated procedure may cause the identification of structures belonging to adjacent acini. This is the phenomenon of "artificial" pores described in the literature, which turns out here to be a digital problem. The problem improves with better resolution, but since artifacts can never be completely avoided, an interactive control is a must for these kind of problems.

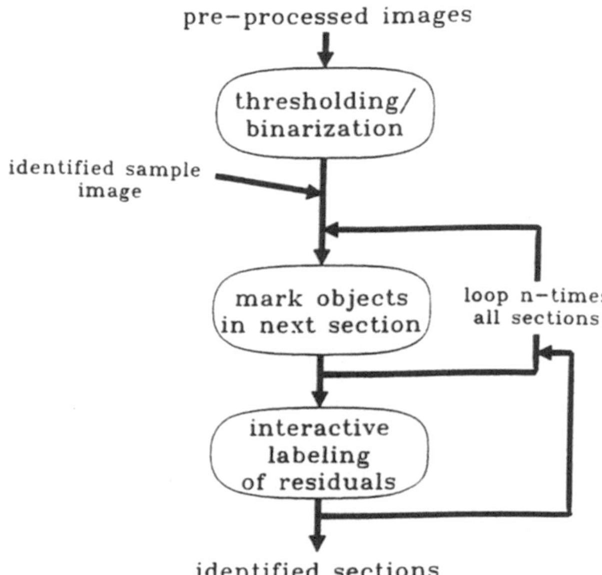

Fig. 14. Automated structure segmentation procedure. The algorithm loops over all sections up and down the data stack. Essentially a mark object operation is used, which identifies a section by the structural knowledge of the preceeding one. Optionally, residuals can be interactively corrected

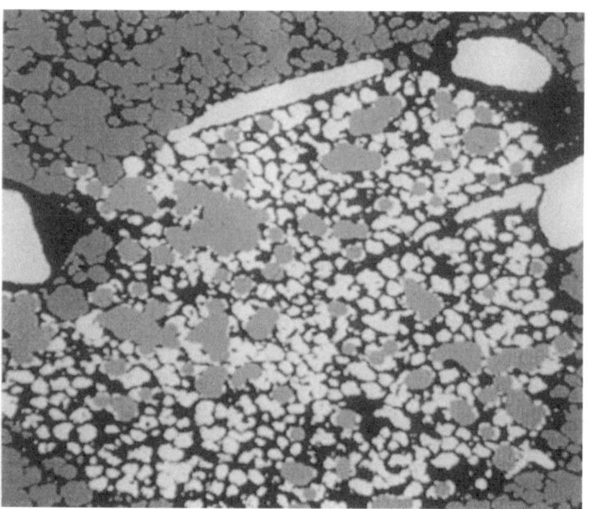

Fig. 15. Example of a lung section, segmented and labeled

3.5
Quantification of Structural Components

3.5.1
Determination of Volumes

After the segmentation was performed, measurements could take place. For the determination of volumes, structures were measured for their areas and all areas of all sections were summed up, scaled, and multiplied by the section thickness. The total volume of the acinus is 0.638 mm^3 (100%). Alveolar ducts have volume of 0.187 mm^3 (29.3%), alveoli occupy a volume of 0.273 mm^3 (42.8%) and the parenchyma left has a volume of 178 mm^3 (27.9%).

The number of alveoli present in the acinus was also of interest. For this purpose the mean volume of an alveolus had to be determined. Area measurements, such as depicted in Fig. 16a, reveal the problem of cut off an only partly represented alveoli in one optical section. As a result the mean of the areas measured is always to small. In addition to applying stereological correction factors a less precise way is to interactively or automatically select structures which fall into a narrower range and such an investigation gives statistics as shown in Fig. 16b.

If we further assume that the alveoli are mostly circular shaped structures, we can calculate their mean volume from the areas measured. The result was 6.348×10^5 μm^3. Dividing the total acinar volume by this size gives approximately 430 alveoli for this particular acinus.

3.5.2
Fractal Analysis of Lung Parenchyma

The measurement of volumes out of serial sections or images is a simple technique. Much more complicated is the determination of surfaces. The surface, unlike the volume, very much depends on the resolution of the measuring device. A related problem is the estimation of the length of object boundaries, which was discussed by Mandelbrot (1967, 1977). As an example, a line has always the topology 1, but its fractal dimension depends very much on its raggedness, which is 1 in case of a straight line, but higher if the line is distorted. This fractal number describes the space-filling ability of a structure, which can have a maximum of 3. The process of converting the quantity length into a dimensionless form is a kind of normalization. For the fractal number, the length or perimeter of an object is related to the reference length or resolution of the measuring device. Plotted in a log–log fashion, the slope of the resulting function gives its fractal dimension (Richardson plot). Inspecting the lung images suggests a fractal investigation, because there might be a relation between the decreasing size of structures and, concomitantly, the number of these structures and their space-filling ability.

Lung sections were investigated at various magnifications. This scaling was achieved by working with different kind of lenses (×2.5, ×5, ×20) and different zoom settings of the laser scan microscope. Images were digitized, preprocessed and segmented and the area occupied by the structures was determined. The smallest

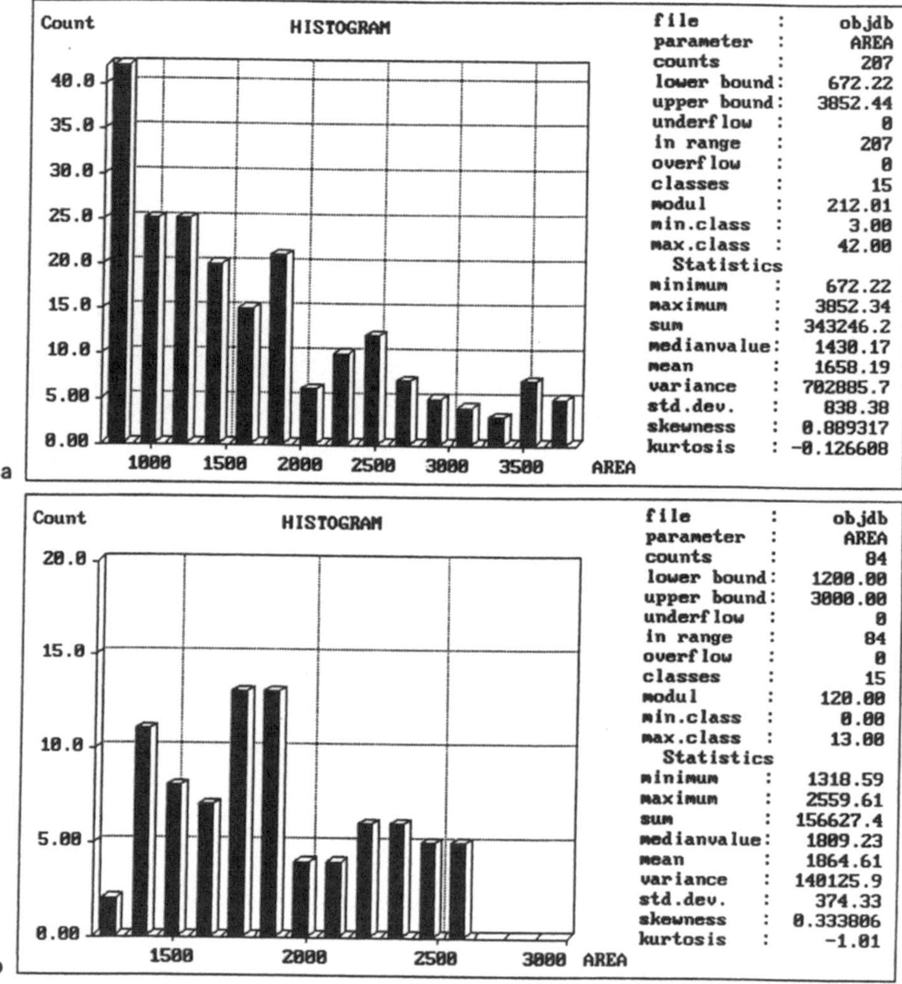

Fig. 16a,b. Area measurements of alveoli in order to determine volumes. Due to the large percentage of cut alveoli the histogram of areas appears biased towards small objects (**a**, area thresholds are 30–150 pixel). Image procesing filters removing these cut objects reveal clearer area distribution of alveoli (**b**, area thresholds are 50–100 pixel)

object measurable after these processing steps was defined as the resolution limit. Unlike similar investigations, structures of the conductive part of the bronchial tree and the respiratory part were handled separately by segmentation and identification. Plotted on a log–log scale (Fig. 17), a difference is seen in the (fractal) behavior of the conductive and respiratory part of lung. The structures of the conductive part of lung fill less space in general but also have a steep slope whilst the structures of the respiratory part fill more space in general but have less slope. Higher slope means better space-filling ability and optimization for minimal work in airflow (Weibel 1993). The design of the ducts in the acini cause a sudden expansion in volume and a slower

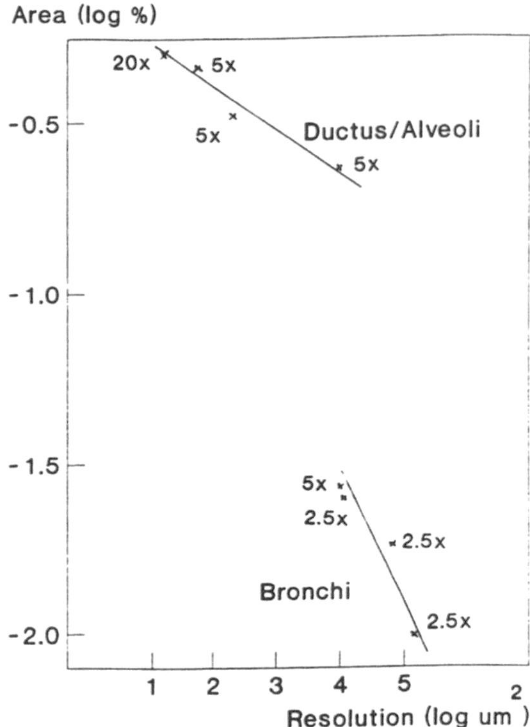

Fig. 17. Log–log plot of the area occupied by alveolar ducts/alveoli and bronchi measured at various resolutions

reduction in airway diameter. Functionally, the air convection is slowed down, giving rise to good conditions for molecular diffusion and uptake. Seen from a morphological point of view, a change in slope of the branching pattern of the bronchial tree towards the alveoli could be interpreted in the sense of some structural limitation.

3.6
3-D Topology as an Analytical Tool

3.6.1
Introduction

The visualization of biological structures based on contour stacks and surface renditions is a commonly used technique, in particular if structures must be outlined interactively for specimens which require a high degree of expert knowledge about the structural composition but cannot be accessed by any kind of automatic image segmentation (Huijsmans et al. 1986; Moss 1992).

The contour-based software described here opens a way to import the structural knowledge of the investigator in order to define 3-D relationships by introducing the issue of topology. The topological treatment of structures has been recognized as a useful way to characterize complex morphological networks (Ley and Pries 1986) and clinical data volumes (Lobregt and Verbeek 1980). In addition to the analysis of branching patterns of lung and neurons (Kriete et al. 1995), other areas of application where this software has been successfully applied to includes dental anatomy (Baumann et al. 1993) and biological structures varying over time (Kriete et al. 1992). The ease of appreciating relationships by graphic 3-D skeletons makes it feasible to investigate complex biological structures and broadens the spectrum of data sets which can be investigated (Kriete and Schwebel 1996).

3.6.2
Basics of 3-D TOP

The software package 3-D TOP is based on contours, which are manually or automatically traced section by section. To define 3-D relationships between contours of adjacent sections, we use the idea of topological lines, which stems from definitions in mathematical stereology (Berry 1976). Figure 19a explains the structure of such a topological skeleton (which is actually a subset of Fig. 3). The branches (B) of a particular skeleton are center lines, connecting the centers of gravity of contours selected by the user. Nodes (N) are defined as points having more than 2 neighbors. Ends (E) are points with exactly one neighbor.

The topological skeleton defines the topology of an object which is mathematically summarized as its genus (G). The genus is determined by G=B-N-E +1 (Russ 1988). For the example given in Fig. 19, the genus is 0. This parameter is independent from the actual shape of the object defined by the contours or the length and form of the branches. It is of great value for many applications to define this kind of property being insensitive to shape and form. Related topological parameters are the number of disconnected parts or objects and their connectivity. An extended pore network may only have one object with high connectivity, whereas a distribution of separate particles without would have zero connectivity and a large number of objects per unit volume.

The topological skeletons are a backbone of the 3-D TOP software because they are not only used to measure topological properties, but they are also used for guiding surface renditions and measurements. In particular, the definition of topological skeletons helps (a) to identify objects and substructures within contour stacks, (b) to resolve ambiguities for surface rendering at bifurcations, and (c) gives way to define parameters for 3-D measurement, in particular parameters being independent of shape.

3.6.3
Modules and Functionality of the 3-D TOP Software

3D-TOP has a total of five processing modules (see Fig. 18). The main features and the functionality of the modules, ordered according to the normal processing sequence, are described in the following.

Module Trace. This module is used to interactively trace contours drawn by mouse or entered on a digitizing tablet. Digitized images may be loaded as background images or may be automatically traced if they are already segmented. The autotracer works on binarized images only and stores the data in a Freeman code format. Figure 19b is an example of a binarized image of lung tissue, taken at low magnification ($\times 5$ magnifying lens). Since the complexity of such a structure is high, the autotracer was used in this case for the generation of contours with areas lying above a defined size.

Editing of the contour files includes delete, add and insert of individual polygons and sections. Up to 1000 objects can be stored in 99 channels, preselected by the user. Channels can be attributed by character labels, identifying their use and nature.

Managing Module. This module allows subselection of sections and channels and definition of attributes for subsequent visualization. Depending on the channels present in the data set, these may be subselected. In this module we have included a section browser, in order to get a quick overview of the data (contours) present in a specific file.

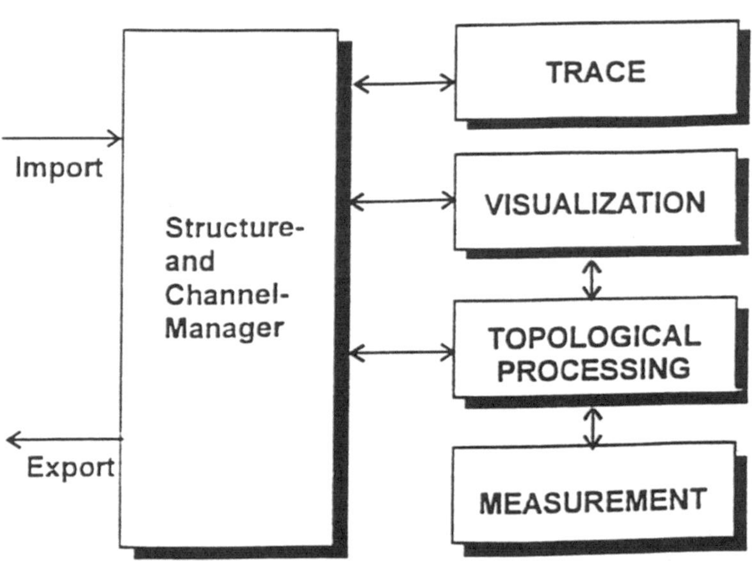

Fig. 18. Processing modules of the 3-D TOP software

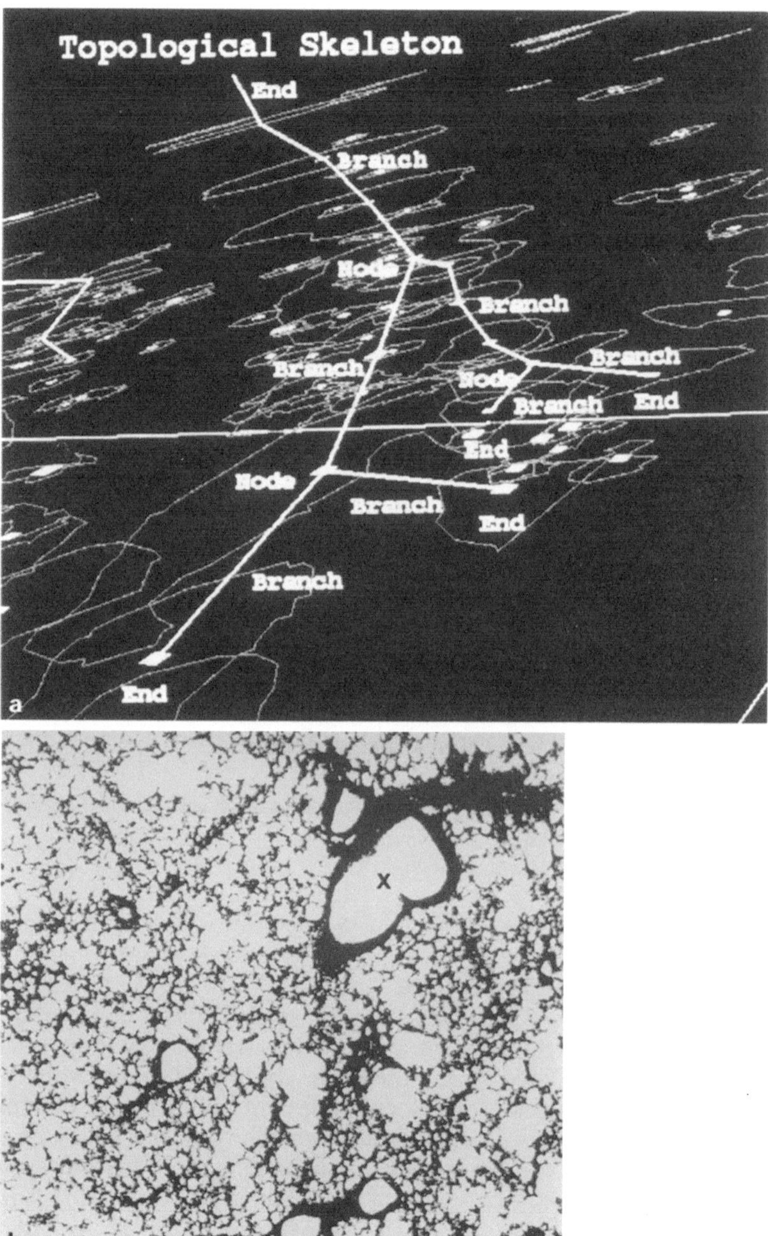

Fig. 19. a A 3-D topological skeleton. The branches (*B*) connect center of gravities (dots) of contours. The skeleton also has nodes (*N*) and endpoints (*E*). This image represents a subvolume of **b, a** binarized image of lung parenchyma (rat). The bifurcation of a bronchius is indicated by an *asterisk*. This particular location is represented by the upper node in **a**

Topological Processing. This module allows definition of the topological skeleton in an exploded view, supported by a semi-automatic generation of topological links. In order to establish these center lines, the centers of gravity are displayed together with the contours. We use the integral of Green to determine the area and coordinates of the center of a particular contour. These centers are linked by bijective correspondence analysis in regions of interest. Bijective correspondence analysis searches for the nearest center of gravity in the following section. In return, it checks whether this newly identified center is also the nearest neighbor to the point where the analysis started. Once confirmed by the user, who can discard any connection, a topological branch is established and displayed. The user scrolls through the data stack back and forth to build up topological center lines. The resulting skeleton can be easily modified by adding new branches or nodes. The software makes it possible for the centers of gravity directly interact in a 3-D view by cursor (hardware picking) in order to edit the skeleton. All branches are stored in an ASCII file, accessible by the software for subsequent visualization and measurement. A typical example of a topological skeleton is given in Fig. 3.

3-D Measurement Module. This module quantifies 2 and 3-D features of the data set, provided that the identification and labeling of a particular structure have been carried out. The user selects the structure at an arbitrary point and a search algorithm identifies all contours which are linked to the selected starting point throughout the data stack. By modification of the topological trajectories arbitrary substructures may be defined and measured.

The contours and topology defined are then subjected to statistical analysis. Parameters present include:
- Individual length, diameter and branching angles and means
- Areas and volumes, sum and means
- Topological parameters such as number of branches B, ends E and nodes N, genus G of the object and connectivity C
- Frequency distribution of branches and nodes.

The genus G of a structure is given by $G=B-E-N+1$. The connectivity is the number of redundant connections in the skeleton, i.e. it describes the number of cuts which can be made without increasing the number of parts.

Visualization Module. This module uses contour stack displays and surface rendering to visualize structures three-dimensionally. The user may select between central and parallel projection. In addition to moving the structure displayed with the menu sliders, the object may be oriented with the mouse controlling tilt and rotation (real-time navigation). When an appropriate view is found, solid modeling and surface rendering based on triangulation can commence. If requested by the user, only those contours are taken into account for rendition which are linked and identified by the topological skeleton.

The triangulation implemented is based on matching a regular geometrical web located around the object onto the polygons. The tiles are filled and shaded. Surface rendering automatically accesses the topological data base, to resolve any ambiguities which may be present. If requested by the user, only those contours are taken into account which are linked and identified by the topological skeleton.

3.6.4
Application to a Lung Data Set

The software was applied to the data sets of lung. Figure 19b shows one section out of a series of 40 sections of the preview. The star indicating a bifurcation of a bronchiolus is represented by a node in the corresponding skeleton at Fig. 19a. Figure 19a is a subset of Fig. 3, representing the whole stack. With the help of topological trajectories, the user can define bronchioli and the main ducts of the acini.

Next, the topological lines were determined for the highly resolved acinar data set. Figure 20 shows a volume rendering combined with the lines, a visualization technique which is described in detail in Chap. 4. Bronchi are displayed in a transparent mode in yellow color, lines in red. The topological skeleton starts with the bronchiolus terminalis, splits into two short segments, which are divided further into two daughter ducts each. These can be easily identified in Fig. 2.

Further analysis of this acinar tree is given in Fig. 21. What is shown is the branching pattern of this tree in a 2-D, projected fashion (Fig. 21a). The arrangement of the plot was designed in a way to give a nonoverlapping view. The length of the horizontal lines correspond to the scaled topological lines. Again, this plot starts with the bronchiolus terminalis. Since the lines are center lines of the ductus and because the x, y, z coordinates are known, the diameters of the acinar pathways were measured from the set of 2-D images and correlated. Since the diameter of the ducts are not shaped evenly along a center line, the mean of the minimal and maximum diameter was taken.

Assuming a circular form of the ducts, the cross-sectional area was calculated. The sum of cross-sectional areas along the distance from the origin of the acinus is given (Fig. 21b), revealing a normal, Gaussian distribution. The maximum of the distribution occurs around the fifth to sixth order of branching. If studied at lower resolution, as for acini A2 and A3 (see Fig. 3), the cross-sectional areas are not only underestimated, but the maximum is shifted to lower distances due to a loss of smaller ducts. Since the volume of these acini is neither known nor measurable, other than these results cannot be reported with confidence.

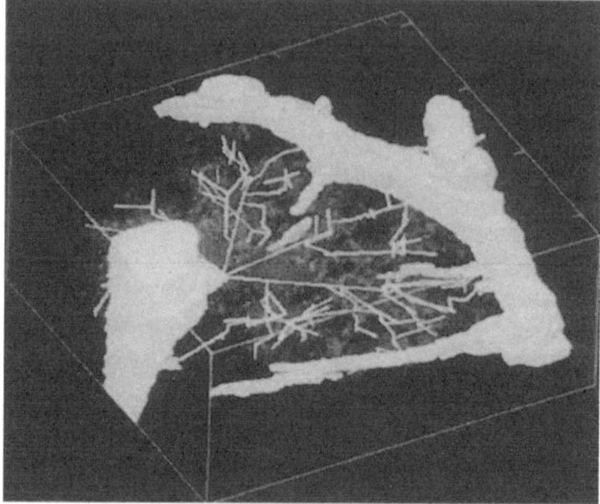

Fig. 20. Topology of acinar ducts

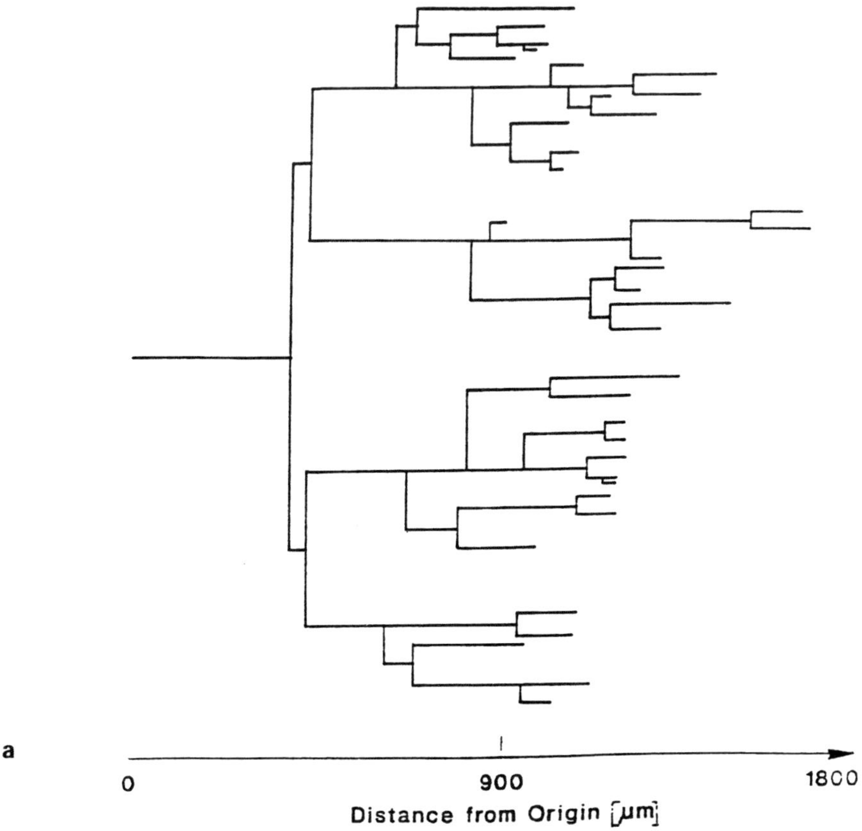

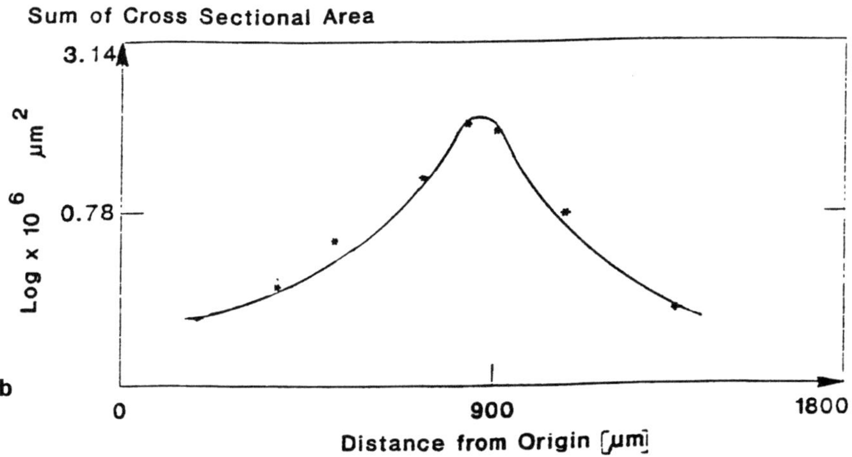

Fig. 21a,b. A 2-dimensional plot of (a) the topological skeleton of the branching pattern of the intraacinar airways and corresponding sum of cross-sectional areas of acinar ducts (b)

38

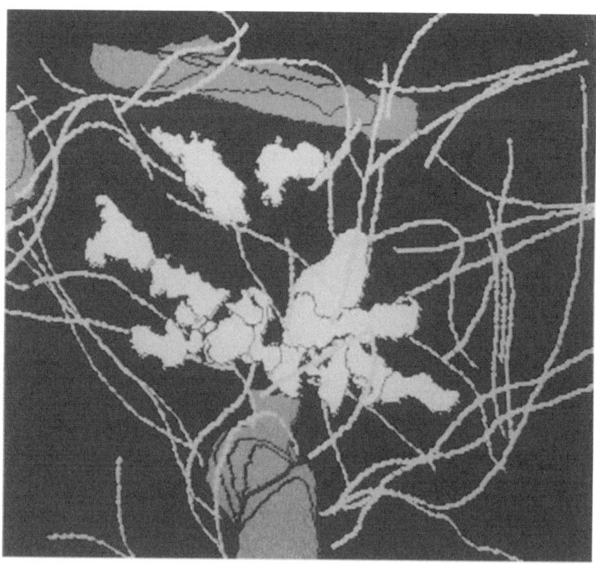

Fig. 22. Contour-based reconstruction in view from top. Interactively, traced septae are added. Image size is 3×3 mm. Six sections represent a total thickness of 420 µm

3.6.5
Analysis of Septae

An interesting structure in the design of lung are septae. These were, however, difficult to detect for the specimen under study. At low magnification and a low contrast setting they most probably showed up, probably by a slightly higher degree of autofluorescence. These septae were manually traced in all sections and reconstructed together with the bronchioli and major ducts (see Fig. 22). Because the spatial behavior of the septae is difficult to recognize in such representations, only views from the top and a selected number of sections are documented here (Fig. 22a, sections 5–8; Fig. 22b, sections 4–9). These septae never appear closed or as polygons and they seem to be related or originate from the bronchioli. They might determine the outer form of the acinus but they also span between the major acinar ducts.

4 3-D Visualization of Microscopic Volumes of Lung

4.1
Basics of 3-D Visualization

Image data of lung tissue, at whatever magnification studied and independent of the means of specimen preparation and image acquisition, are dense and extremely complex in nature. Simply displaying such a 3-D data set graphically on the computer screen does not provide any insight or comprehension. The purpose of 3-D visualization is to transform such data into a visual form that promotes perception and understanding.

The use of computers for the purpose of 3-D visualization in biomedicine was reviewed in several articles previously (Katz and Levinthal 1974; Ware and LoPresti 1975; Bloch and Udupa 1983; Huijsmans et al. 1986; Kriete 1992; Hersh 1990). In microscopy, sophisticated methods and algorithms to align mechanical serial sections, such as in histology or electron microscopy, are available now. Together with the realization of the confocal principle in optical microscopy, the methods to improve the axial resolution in fluorescent microscopy and the recent advances in NMR microscopyform a class of acquisition techniques which fulfill the necessary requirements for volumetric data representation purposes. Otherwise, misalignment and sampling errors become quite obvious in volume rendering, in particular if the thickness of the multiplanar sequences must be adjusted to the lateral resolution of the voxels, thus exhibiting cubelike elements. Isotropic sampling guarantees for the representation of the volumes with a high degree of volumetric fidelity (see Appendix 1).

The software available to visualize 3-D microscopic data reflects the two worlds of imaging and graphics (see Fig. 23). This division is based on the fundamental difference between the elemental data used, either a geometric entity (polygon) or numeric entity (pixel). Data are fed into the computer by tracing continuous contours or by digitizing images into discrete matrices. Accumulation of the 3-D information is performed by stacking up the individual sections.

Contour extraction is a form of data reduction and was the preferred technique when computer power was rare. In contour representations, objects may be formed by a tiling procedure and the fundamental surface element in 3-D is a facet. In pixel-based processing the fundamental entity of 3-D imaging is a volumetric element (voxel). Consequently, the techniques to obtain a 3-D reconstruction of multiplanar images can also be subdivided into two main groups: (a) algorithms which perform surface reconstructions based on contours and (b) algorithms which represent volumes directly, a technique called volume rendering.

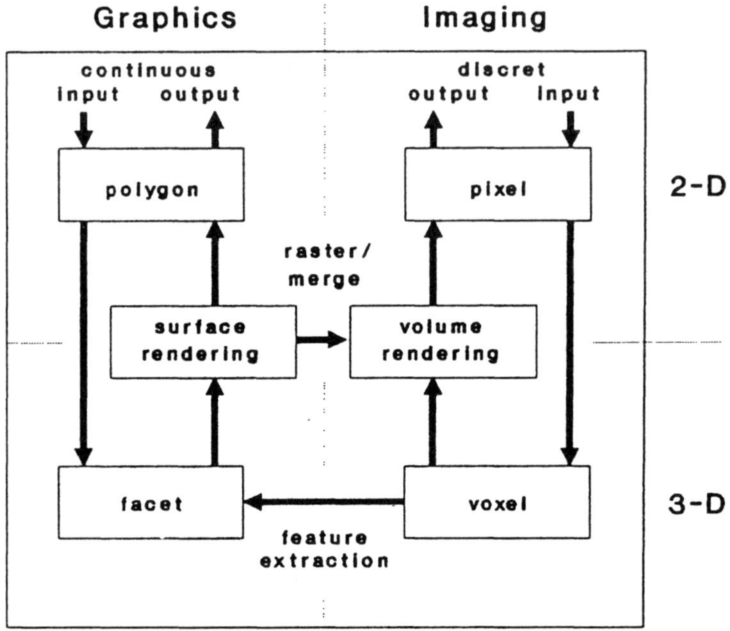

Fig. 23. Basic entities used in 2-D and 3-D imaging: graphics and imaging vs two and three dimensions (from Kriete, 1992)

Contour extraction is a form of data reduction and was the preferred technique when computer power was rare. Surface reconstructions based on contours presume that the data volume is populated with structures of distinct boundaries whose shape is known a priori or can be accurately determined. Errors in the alignment or differences in resolution can be substantially suppressed by triangulation, shading and coloring. When this happens, the fidelity of the reconstruction becomes questionable.

Volumetric data representations have their roots in digital imaging. Images are stored in the computer as 2-D matrices of picture elements, and properties of the objects imaged are coded by the digital gray level values. The discrete nature of digital images has led to a number of theories which are associated with the famous names of Nyquist and Fourier. In 3-D imaging typically sequential 2-D images are assembled, which extend the image matrices into depth. The resulting volumetric elements or "voxels" are the scalar primitives of volume representations. This is unlike to conventional 3-D computer graphics, where geometric primitives such as polygons are used for subsequent tiling and surface display. To transfer volumetric representations into a visualization which allows a perception of the spatial object arrangements, a process is performed called volumetric rendering, taking into account the voxel values.

Another feature of volume displays, often combined with volume rendering, is in the use of 3-D cutting functions, which can be performed under interactive control and in real-time if advanced hardware structures are used. Cutting functions have the purpose of representing planes other than the image sequence planes. A special class of such functions first reconstructs the object under investigation at lower resolution to interactively orient the image stack. Hereafter, a surface view with full resolution is calculated and a cutting plane can be interactively shifted and rotated.

In view of the distinct advantages of volume rendering over surface rendering for a dense, complex data set of an acinus, various types of volume rendering were investigated and applied. For all applications, the Voxel View Ultra Software (Vital Images) and the Mipron Software (Kontron) was used in the following.

4.2
Types of Volume Rendering

Volumetric representations include the cuberille, the octree and the voxel. First attempts to model volumes have their origin in the cuberille representation (Herman and Liu 1979; Herman 1981; Herman and Udupa 1983). The cuberille starts with a segmentation process to extract the voxels of interest followed by a description of the surface of these voxels. These voxel faces represent cuberilles which may be rendered. The data reduction inherent in cuberille representations has made this kind of processing feasible when computer power was rare. To date, no direct application of this kind of processing has been reported in the field of microscopy and therefore the cuberille is not considered for a detailed discussion in this chapter.

Another way of volumetric data representation is octree encoding. This computation recursively subdivides the data down to the finest resolution available in a tree-like fashion. Subdivision of volumes and subvolumes is only performed if differences in intensity at a certain level exist, and each lower level increases the number of elements by a factor of eight. Octree representations together with an inherent classification may reduce volumetric data enormously and speed up visualization and interactive operations (Smith 1989; Kriete and Pepping 1992). In addition, the solid volume defined can be used for morphometric evaluations. Octrees are less commonly used in microscopy, but they have the potentials of data reduction, resolution dependent analysis and parallelization. This kind of coding and transformation of data have also been realized in so-called wavelet transforms (Westermann 1994), and rendering algorithms are being developed to directly work on this kind of compressed data, so that large volumes can be visualized with small workstations. Typically, image sequences require 16–256 megabytes of computer memory.

A "brute-force" 3-D reconstruction, a task only feasible with advanced hardware structures, accesses all voxels within a volume directly. This means of voxel processing, often directly combined with rendering, has led to an acceptance of the general term "volume rendering" for this class of algorithms (strictly speaking, volume rendering also holds for the visualization of other volume representations).

4.2.1
Image and Object Order Rendering

Voxel representations have the advantage of being free of any prior knowledge of the data set. This direct way of processing works without preprocessing or segmentation. However, a number of applications include a preprocessing step to especially visualize volumetric attributes and sub-structures. Unlike to "analytical," contour-based reconstruction techniques, complicated objects with free forms and fuzzy boundaries can be visualized directly in volume rendering. The algorithms project rays through the volume following a user-defined geometry and the different attributes of the voxels such as intensity, color, opacity and their gradients can be taken into account. 3-D volumes can be reconstructed at any orientation and interactive slices through the volume can be performed.

4.2.2
Implementations

Figure 24 gives an overview of specifications to describe volume rendering algorithms. The different features have been compiled in a way that the two most commonly used implementations, image order and object order rendering, are reflected on both sides of the table. In principle, other combinations or variations are possible.

One method used in volume rendering is to define imaginary rays through the volume, so called ray-casting techniques (Roth 1982; Tuy and Tuy 1984). The way of casting is accomplished by the definition of certain projection geometry. Assuming a central projection, the angles of the rays are defined by the matrix of the screen and the center of projection (Fig. 25a). A common implementation traverses the data in a front-to-back order, which has the advantage that the algorithm can stop when a predefined property of the accumulation process has been reached. This class of methods is also referred to as *image order rendering*, since the matrix of the screen is used as the starting point for the accumulation process. It is obvious that such a ray-firing technique does not necessarily pierce onto the centers of voxels, as illustrated on the cross-section (Fig. 25b). An interpolating process must take into account adjacent voxel values, which makes the rendering process slightly more complicated. Image processing systems are well suited to implement ray-casting routines.

A much simpler geometry is realized in a parallel projection, as illustrated in Fig. 25c,d. In this approach, also referred to as *object order rendering*, the volume is rotated according to the current viewing angle and then traversed in a back-to-front (BTF) order. During traversal each voxel is projected to the appropriate screen pixel where the contributions are summed or blended, a situation which is referred to as *compositing*. As shown previously (Gordon et al. 1985), such a class of algorithms allows an efficient readout, and due to the simple geometry computation times are much lower in object order rendering than in image order rendering. Some of the requirements of compositing voxel attributes make graphics hardware an ideal platform for implementation. One recent example is the use of texture elements and texture computer buffer giving superior volume rendering performance such as displaying a data set of 256×256×256 voxel in real-time (Cabral et al. 1994).

On the other hand, image order algorithms have some advantages over object order algorithms, in particular due to the use of a perspective projection giving rise to very natural views. Since the resolution of the display is the limiting factor, and not the voxel primitives of the data set, smooth zooms can also be realized.

Method	Image order	Object order
Geometry:		
Type of projection	Perspective	Orthogonal
Procedure:		
Traversal scheme	Ray casting (front-to-back)	Back-to-front (compositing)
Voxel representation	Intensities (binary, gray level)	Attributes (opacity, color)
Implementation:		
Display table	Gray/color LUT	Alpha LUT
Hardware platform	Imaging	Graphics

Fig. 24. Methods and specifications used in volume rendering

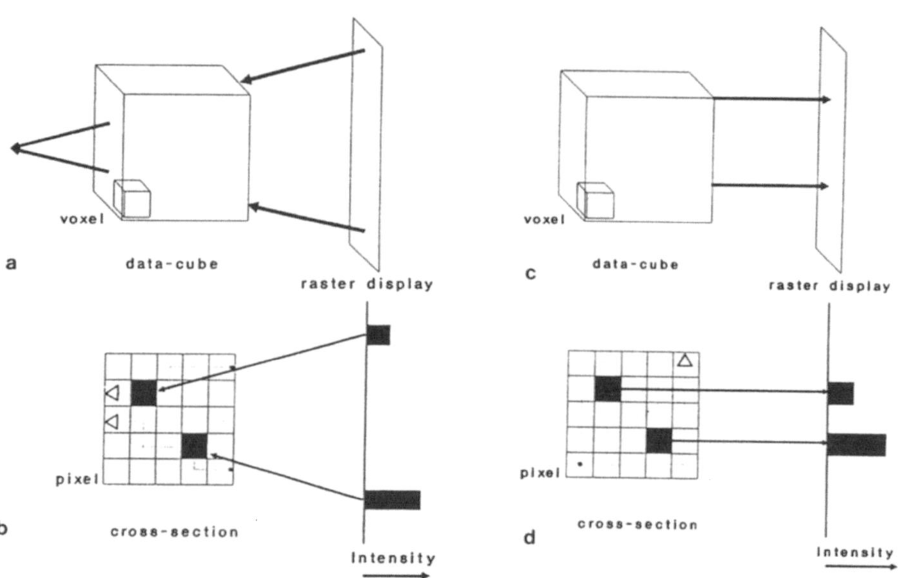

Fig. 25a–d. Illustration of image order rendering (**a,b**) and object order rendering (**c,d**) (from Kriete and Pepping, 1992)

4.2.3
Voxel Intensities and Image Order Rendering

Image order rendering accumulates the voxel intensities directly. The most frequently used modalities in ray-casting include the detection of maxima, the integration of voxel values and the detection of voxel surfaces (Hoehne et al. 1987; Robb and Barillot 1989).

The first ray-casting modality to be applied to the acinus data set is called the *maximum mode*. The maximum gray level along a ray traced, lying above a user defined threshold, is visualized. The procedure allows some kind of transparent visualization of bright structures under a surface. This is, however, not the case for the uniformly

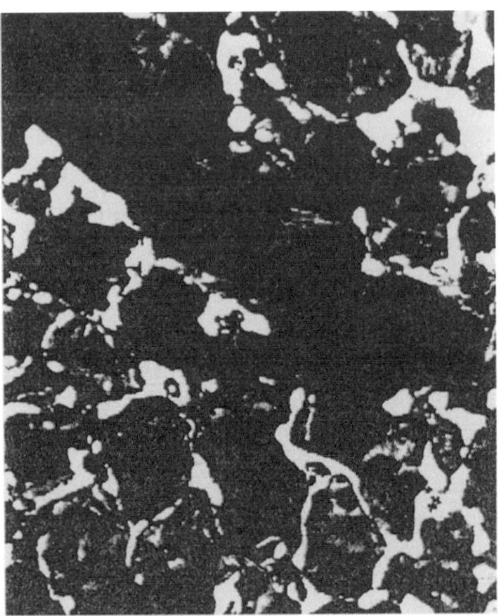

Fig. 26. Image order rendering of a lung tissue. The metallic looking appearence is caused by a gradient shading

structured lung tissue. Another way of ray-casting visualization is the *integration mode.* The effect of integrating voxels did not work out for lung specimen because of the dense and complex structures. The higher the integration values are, the more fuzzy the 3-D reconstructions appear.

The most simple, but also most sensitive implementation of the ray-casting algorithm is the *surface mode,* a display combined with different shading algorithms. Thresholding the gray level images and subsequent binary image processing such as opening or scrap allows a further clarification of the image content. The class of such volume rendering procedures is referred to as binary volume rendering (Fuchs et al. 1989). This turns out to be very useful for lung structures and an example is given in Fig. 26. The gradient shading which was applied gives this image a high contrast and shining appearance.

None of the above algorithms applied to the lung structure, however, provide the necessary insight required; in particular there is a lack of differentiation of the individual classes of objects present in the data set.

4.3
Voxel Attributes and Object Order Rendering

Implementations of volume rendering try to enhance the visualization in general and in particular of interesting volumetric substructures. Since the appearance of a point on the screen depends on the properties of voxels during the accumulating process, other parameters besides the intensities may be considered.

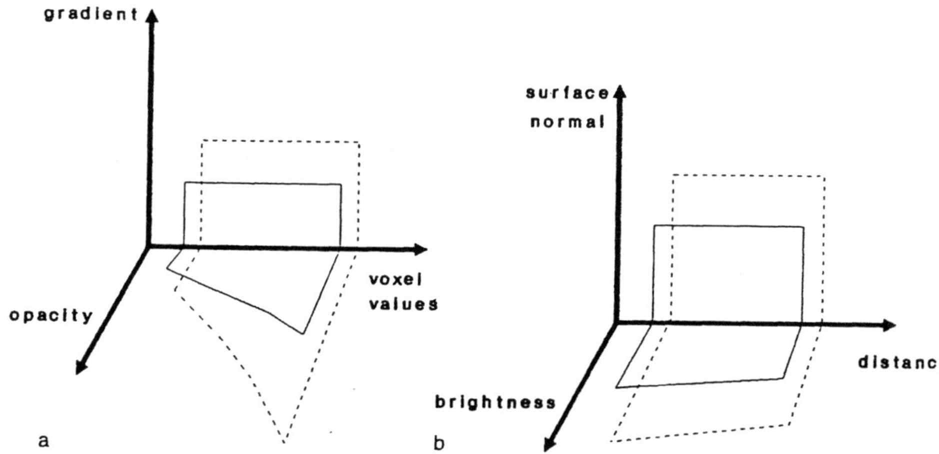

Fig. 27a,b. Application of additional lightning modalities in object order rendering. **a** Calculation of voxel opacities for two different gradients in a range of voxel values. **b** Calculation of voxel brightness for two different surface normals in a range of object distances (from Kriete and Pepping, 1992)

Typically, opacity and brightness are such attributes beyond color, location or gradient values. The rendering with the activation of such attributes can be done in a separate preprocessing step or can be directly linked to the accumulation process (Debrin et al. 1988; Levoy 1988), such as in object order rendering.

For a better description, we first assume a separate preprocessing step, this means a new scene S^n is calculated out of the original volume V_n (Udupa and Hung 1990). The *opacity* of a scene determines the light transmission property. The tuning of the opacity being dependent on the voxel values is demonstrated for the preview data set in Fig. 27. In Fig. 27a the opacity is generally high for all gray levels. This prevents the recognition of bronchial structures with a high gray value, because all other structures with lower gray values prevent a looking-through effect by the high opacity applied. Reducing the opacity, or improving the transparency of the tissue, clarifies the image as depicted in Fig. 27b.

Also the vicinity of a certain voxel value may be considered, in order to enhance boundaries with high opacity values. Voxel values can be assigned to opacity values in a linear or unilinear relation. In contrast to integration, where all voxel values have equal weight, such procedures allow for a voxel value dependent weighting of structural properties.

As an example, an opacity scene S^n is defined as $S^n=(V^n,f_d)$, whereas f_d is a linear function of density and the magnitude of the gradient. Figure 28a demonstrates two different gradients over a range of voxel values and the resulting magnitudes for opacity.

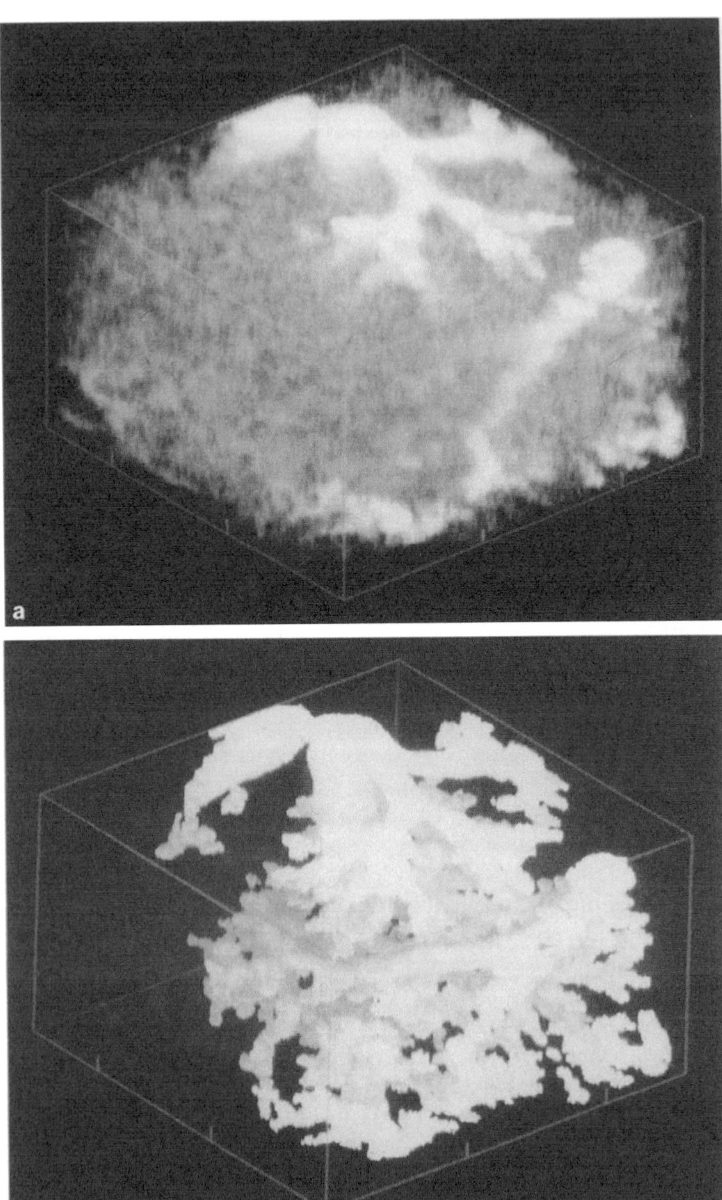

Fig. 28a,b. Preview data set in volume rendering mode (compare with Fig. 3). High opacity (**a**) and low opacity (**b**) allow tuning of the appearance of the airways

4.3.1
Alpha-Blending

If the rendering of voxel attributes is combined with the idea of object order composition and implemented on a computer graphics display, volume rendering is used with the idea of *alpha-blending*. For this purpose, a so called look-up table (LUT) is used, which relates the intensities of the objects to be displayed to colors. For volume rendering, the alpha LUT or opacity/transparency LUT is the most powerful table. It is this table which governs the blending process, allowing one to control the opacity of the voxels in a volume and hence see voxels near the faces of the volume and voxels deep inside a volume at the same time. In its simplest implementation, the alpha LUT is a 1D map which maps the range of voxel values (0–255) into a range of so-called alpha values (0–1.0).

The alpha value of each voxel determines how that voxel is summed, or blended, into the screen pixel onto which it is projected. The standard blending equation for back-to-front compositing is generally as follows (Eq. 4, VanZandt 1991):

$$\text{Color(out)} = \alpha(v_i)\,\text{Color}(v_i) + \text{Color(in)} * [1 - \alpha(\text{in})] \tag{4}$$

In this equation *Color* is a vector composed of three color display components R, G and B; the alpha value of the new voxel being projected to the screen is used as a weighting factor to determine the effect that voxel will have on the current color of a pixel. For example, a voxel with a high value will be more opaque than a voxel with a low alpha value. When such a voxel is projected to a screen pixel, its color will tend to dominate over the existing color of the pixel. As this procedure is applied recursively for all the voxels in the volume, it can create the visual effect that one can see through voxels of low opacity in the foreground and thereby see voxels further away.

Figure 29 gives an example of the preview data set, where the segmented structures (bronchiolus, major acinar pathways) have been assigned to colors. The opacity was set to the maximum value for all structures displayed. Applied to the segmented acinus data set, different structures are displayed with different opacities and colors. Figure 30 has been rendered with balanced opacity values. Bronchi are given in yellow with shading to give additional brightness. Ducts are displayed in red with mean opacity. Alveoli are displayed in green with low opacity. The berrylike appearance of these structures could be very much enhanced by a special lightning technique described below. This visualization uses the embedded geometry feature to give an analytical view of the branching pattern, as explained in Sect. 4.4..

4.3.2
Brightness

The *brightness* of a voxel if rendered to the viewpoint is of special importance to improve the visualization and spatial impression. For the data set under study, the great amount of alveoli gives rise to some irritations in some parts of the image; in particular it is difficult to see their relation to other structures. The main idea of assigning brightness to a scene is to use shading methods which have proven successful in surface rendering and binary volume rendering. Hereby, a brightness scene

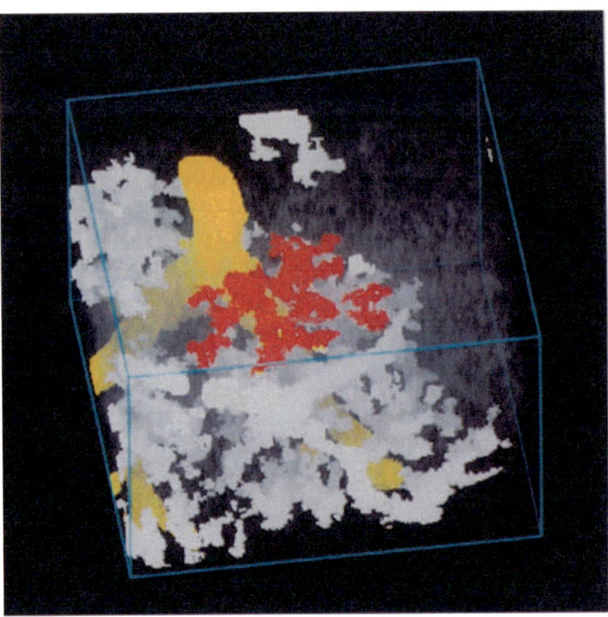

Fig. 29. Same preview data set as in Fig. 28. Here the label of structures is considered and visualized as colored forms and high opacity. Main duets are displayed in red, bronchioli in yellow and undifferentiated pathways in white

$S^n=(V^n,f_b)$ is computed with a function f_b which assigns brightness values to each voxel. This is described with the help of Fig. 27b. Herewith the brightness depends not only on the distance of the voxel to the viewpoint but, in addition, on a normal vector of the voxel surface with respect to the voxel neighbors.

The final rendering of these scenes has the objective of creating a view taking into account a combined process of reflection from and transmission through the voxels along the ray fired (Udupa and Huy 1990). Assuming a brightness value b(in) ambient to the light intensity, this value is modified while traveling through a certain voxel v_i toward the viewpoint resulting in b(out) (Eq.5):

$$b(out)=f_o(v_i)\, f_b(v_i) + b(in)[1-f_o(v_i)] \tag{5}$$

where $f_o(v_i)$ and $f_b(v_i)$ are the opacity and brightness values of this particular voxel. The b(out) value is computed by a linear combination of the opacity weighted by the brightness scene values and the incoming brightness weighted by the opacity.

Different shading methods for volume rendering exist. The algorithms show different ways in approximating a curved surface from polygons or voxel boundaries and different formulas for the assignment of a shade value to a particular pixel in the picture plane. The three grounds on which different techniques can be compared are applicability, image quality and computation effort.

In some simple shading models the shade of a particular point on a surface depends on some local surface attributes without considering the context of that point (noncontextual shading).

50

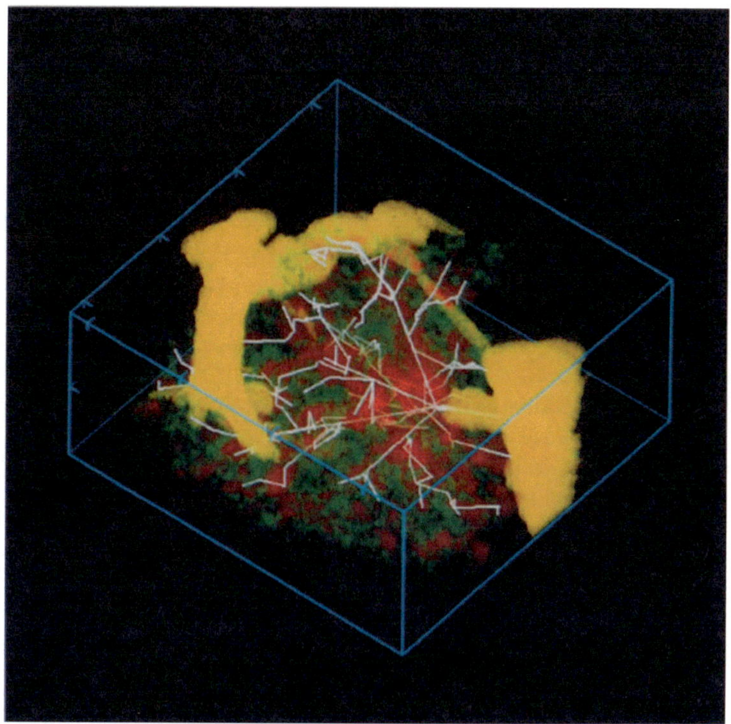

Fig. 30. Volume rendering of the acinar data set with respect to segmentation and label. *Yellow,* bronchioli; *red,* acinar pathways; *green,* alveoli. Tuning the gray level/opacity relation allows for a balanced appearance. An additional gradient shading of the voxels enhances alveoli and bronchioli

In image space shading an attractive implementation because of simplicity and speed is distance-only-shading. The calculation of intensity depends only on the distance d of a point P at the surface to the viewpoint. Starting with an offset of c, the distance gives the new value I(P). In a modification, another method uses the angle phi between the surface normal S and the incident light, called constant shading.

Distance-only-shading can lack in structural perception. A substantial improvement is in the application of a gradient to any shading generated, but in particular to distance shading (see Fig. 26). Such gradient-shading (Gordon and Reynolds 1985) improves the recognition of details at slightly inferior image quality than object space shading methods discussed below. Since the real surface often is approximated by polygons, geometric solids or voxels, those pictures contain artifacts such as sharp edges of the voxels. That is why those simple shadings are not satisfactory in times of increasing computer power.

Especially for volumetric data representation such as cuberille and voxel, a context-sensitive shading called "normal based contextual shading" was developed by Chen et al. (1985). The method profits by the property of voxel models in which the approximated surface consists of single squares that are parallel or perpendicular to each other. The advantage of the method against noncontextual shading is that the generated pictures look smoother and are invariant against rotation. Now as before the appearance of artificial voxel- or solid-edges is disadvantageous. To increase the realism of a shaded picture, there are two well-established interpolation methods developed by Gouraud (1971) and Phong (1975). While the Gouraud shading interpolates shade values across surfaces, Phong's algorithm interpolates the normal vectors before computing the resulting shade value.

A different method specifically for computing especially pleasing visualizations is in ray tracing (Goldstein and Nagel 1971; Rubin and Whitted 1980). Ray-tracing, a special type of shading routine, should be further differentiated from ray casting, a volume rendering technique. In the ray-tracing model the rays are sent from the light source to the data-cube where they are reflected, absorbed or refracted by the objects. The rays may be computed further by means of geometrical optics to illuminate other objects in the scene or to cast shadows. Consequently, the main difference above other shading algorithms is in the possibility to go beyond pure surface representations. This method can take much more expense for computing a single ray than ray-casting, but can generate very natural views. Parallel computer architectures can help to speed up ray-tracing operations (Cleary et al. 1986).

The effects of applying a lightning model are highly data dependent. For data which contain objects having smooth surfaces or smoothly varying contributions, the rendered images can be nearly indistinguishable from ray cast images. On the other hand, for very noisy data, the application of the lightning model can create images in which the shading merely serves to amplify the noise. The data set of the acinus is preprocessed in way that the noise is widely filtered out. After assigning brightness, the sphere-like acini reflect in a similar fashion if illuminated from one direction. By that, the perception of their spatial arrangements is enhanced.

4.4
3-D Imaging Meets 3-D Graphics

The philosophy of volume rendering and its ability to render noisy, fuzzy scenes and structures without a-priori knowledge can be shortcomings at the same time, in particular if volumetric structures must be described quantitatively. Such a task can be easily achieved in analytical, contour based approaches. But also for advanced visual renderings, modeling and simulations the shortcomings of volume rendering are obvious. To overcome such limitations in volume rendering, it seems to be desirable to implement computer graphic elements. Such an implementation does not only include a software problem, but any solutions have to take into account the restrictions of the computer hardware.

One attempt is where polygons are extracted out of a volumetric data set. A more extensive attempt was realized by the implementation of a 3-D rendering on a computer graphics platform (Forsgren 1990). In conjunction with 3-D filtering techniques surfaces of classified objects in a volume are extracted and the resulting subset of voxels is stored as vectors together with the voxel values. This coding exhibits a strong data reduction and enables an easy implementation on graphic processors, taking advantage of a high rendering and visualization speed. However, if different aspects of the volume must be enhanced, the vector description must be calculated anew.

Another link between voxel representation and graphics is in the integration of simple geometrical features into the process of visualization. Such geometrical attributes expressing local properties of the structure can be either interactively or automatically extracted, or be added as modeled features (Levoy et al. 1990). In order to allow an unhindered access to such attributes for the purpose of interaction to the data, such as rotation and cutting, such features can be directly integrated into the volumetric data set, or are handled as an overlay by using alpha-blending. Another way is in the merge of such features in a buffer, such as the Z-buffer.

Examples for such an embedded geometry are given in Fig. 30. These lines have been imported to the EmbeddedGeometry feature (as part of the Voxel View software), which allows geometry-based and voxel-based rendering to be merged into the same image. After importing, some of the major lines have been supplied with thickness. These lines support the structural recognition, in particular that of the ductus.

Extensions of this technique concerning complicated objects and their attributes may be found in a separate volume and surface rendering and subsequent software merging (Fruehauf 1991). As an example, ray-casting rendering can be combined with ray-tracing surface rendering, because the two techniques feature similar geometries. As a result, the output of both renderers is stored as a depth-sorted list of so-called image space elements (ISELs), which can be further reduced and merged. Such a method has a great potential for being implemented in parallel architectures.

4.5
Stereoscopic Displays and Virtual Reality

Despite shading and rotation, computer reconstructions are 2-D displays. For a long time stereoscopic displays have been tested. Basically, we can differentiate between time-multiplexed and time-parallel systems (Hodges 1992). Time-multiplexed are mechanical shutters or active crystal glasses synchronized to the refresh frequency of the monitor. Time-parallel systems include red/green displays, dual screens or head-mounted displays. Wearing a head-mounted display (HMD) gives the user the sensation of being immersed in a 3-D, computer-simulated world. We have therefore tested the possibility to use VR techniques for the visualization of the microanatomy of lung, a technique which could be named micro-cyberspace.

It is essential to notice that VR techniques require very fast computers – even supercomputers and graphic supercomputers if available. The reason is that the user moves the display or his position and orientation is tracked, and each movement requires that a new sight be calculated. Delays in calculating views are unsatisfactory or might cause sickness, because what the eyes see does not naturally fit to what is expected by the movement of the body or the head. Moreover, two images, for the left and right eye, must be calculated with the right perspective.

We have used a graphic Supercomputer Onyx (Silicon Graphics) with the Reality Engine 2 at GMD, Bonn, to render a surface display of lung tissue. Two VR display devices, linked to the Onyx, could be tested: a head-mounted equipment from VPL and the BOOM from Fake Space Labs (see Fig. 31). The necessary surface elements were calculated with a program running on a Connection Machine 5 supercomputer, featuring 64 processors. Surface tiles were transferred by HIPPI fast network link to the Onyx. At a data set of 64×64×64 pixel resolution over 25,000 tiles were generated

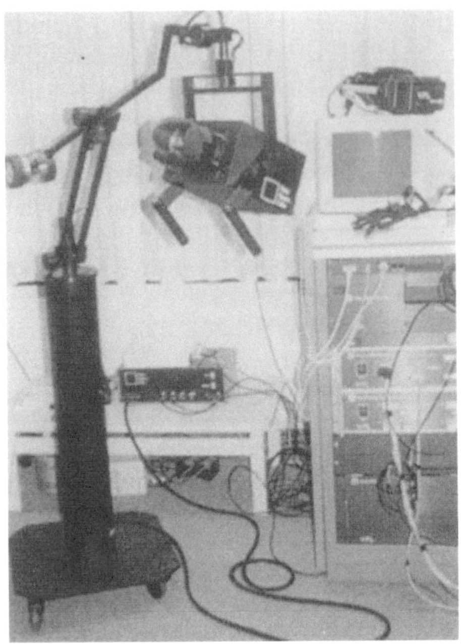

Fig. 31. Virtual reality equipment: BOOM (*left*) and head-mounted display with data gloves (*right*)

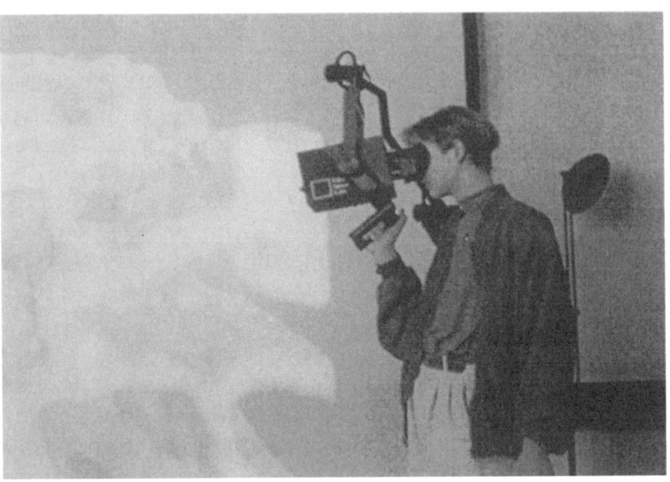

Fig. 32. Use of the BOOM to inspect a lung data set also visible on the screen

due to the complexity of the lung structure. This is an extremely high number compared to usual VR applications such as design, robotics, etc. (Emmet 1992). Consequently, the refresh of the display was in the range of 1–3 s for this kind of data when tested (see Fig. 32), although it was instructive to inspect the alveoli and their arrangement with this technique. What is requested from computational speed would be in the range of 1/60 s per frame. This shows that right now these kinds of systems can only be applied to very simple biological forms, but that the next or subsequent computer generation might be powerful enough to begin applications in biomedical imaging, in particular in microscopy, on a broader scale.

5 Discussion of 3-D Analysis at Respiratory Units

Scientific computing has been offered the challenge of becoming as an important research tool to lung research. Scientific computing is understood here as a sophisticated technique for acquiring a large, highly resolved data set representing a complete acinus and to analyze and to visualize it, in order to see into and to comprehend complex lung structures.

So far investigations of the acinus rely on cast techniques, which make it possible to analyze the outer form of the acinus or to quantify global parameters such as diameter or volume. The internal structure remains hidden in these preparations. Serial mechanical sectioning in conjunction with traditional microscopic techniques delivers only 2-D insights of a structure which is essentially 3-D. The reasons which have prevented a highly resolved, 3-D analysis of an acinus are the need for imaging quite a large cube and a high resolution to resolve structural details. Since no imaging technique is available as of yet which combines both these requirements, the only alternative is to cut sections but the disadvantages at the same time are the fairly high number of sections required which govern the axial (z) resolution and the need for precise alignment before any further digital processing can commence. If these requirements cannot be fulfilled, the image interpretation can fail due to the occurrence of artificial openings (pores) in the lung parenchyma.

A new microscopic imaging technique presented here is confocal microscopy. The main advantage over usual light microscopy is its improved axial resolution. In addition to instrumental, physical and staining problems, image quality aspects are examined. Discussions of image quality must include various constraints and conditions. In general we find that the image quality must first be considered as noise limited. If we can create a situation of noise free imaging, then the diffraction-limited parameters and transfer functions define the achievable performance of an imaging device. Image quality criteria take into account these conditions and must be chosen properly. In fact, there are no universal and unique criteria, but we may choose from a set of available methods to measure different properties such as sharpness, contrast and fidelity. We have done some initial attempts to describe image quality three-dimensionally, which is useful for the design of the 3-D imaging apparatus such as the confocal microscope, the optimization of the setup, as well as for the subsequent digital restoration of the sampled data (Kriete 1994). One key term is volume fidelity, to describe the similarity of an object and its volumetric image. We have seen that the improvement in image quality of confocal imaging over conventional imaging is a measurable quantity and future work in this area must be encouraged.

Confocal microscopy is the key technology to assemble a large microscopic volume. The method developed here uses a computerized, large area scanning technique applied to thick serial sections, which are resolved axially by optical tomography. This guarantees for an extended field of view, a high resolution laterally and axially and a reduced workload for image alignment and hence less risk for generating disconnected structures.

H&E-stained sections of 70- to 80-µm thickness are observed in (auto)fluorescence. A framework during data acquisition is used for reproducible scanning. This method includes the definition of a region-of-interest of a preview data set scanned at lower resolution. Moreover, the preview data set is aligned and transfered back to the microscope for alignment of sections which are scanned anew at higher resolution. A mosaic of areas is taken by electronic beam shifting. This reference frame of nine images serves for identifying areas which are scanned with a higher numerical aperture lens and which resolve the sections axially. The nine data cubes are fused at every level of optical sectioning and combined with the data contributing from 12 other histological sections. The resulting data set has a field of view of 1800×1800×910 µm and a digital resolution of 768×768×91 voxel and covers the space of an acinus completely. The framework is also essential to correct for errors which may occur in image uptake or storage and almost allows us to find the right areas in the specimen.

The data set was digitally processed and first preprocessing and segmentation carried out. A recursive segmentation algorithm was tested successfully to mark objects of adjacent sections, which reduces the interactive workload for this kind of complex data set. Interactive correction of the labeled structures was nevertheless still required. Subsequently, the volumes were quantified. The volume of acini reported varies for rat between 0.53 mm³ (Mercer et al. 1991), 1.86 mm³ (Rodriguez et al. 1987) and 10.27 mm³ (Valerius 1988). Here the volume is of 0.63 mm³; this is closer to the results of Mercer, but it is must be clear that there is a variation between the acini and probably due to their location and that the way of inflation/preparation is also something to be considered. The measurements of the mean volume of human acini reported also differ remarkably between 15.6 mm³ (Boyden 1971), 182.8 (Hansen and Ampaya 1975) to 187 mm³ (Haefeli-Bleuer and Weibel 1988), and the number of acini for humans differs between 15,0000 (Weibel and Taylor 1988) and 30,000 (Haefeli-Bleuer and Weibel 1988) correspondingly. In any case these investigators found also variation of the size of acini within one organ, which might been important for the function of the respiratory units.

Structural components of the acinar volume measured separately include acinar pathways (29.3%), alveoli (42.8%) and parenchyma (27.9%). The numbers which result from such an analysis, in particular that of the parenchyma, very much depend on the acquisition parameters and the quality of image processing. During the acquisition in the LSM a blooming of structures was observed, due to the autofluorescence and the need to find a proper adjustment of gray levels by contrast and brightness. Structures lying beyond the resolution of the instrument might have also been detected. Subsequently, the size of the digital filters and the way of thresholding might influence the results so that the volume of the parenchyma might be actually less than reported here.

A topology software which has been developed to meet the needs of those who analyze mechanical serial sections in histology and optical sections in laser scanning microscopy was also applied to this data set. The semi-automatic definition of topological center lines was the backbone of this software. The three basic advantages in using the concept of topology are (a) resolving ambiguities for surface rendering,

(b) flexible definition of objects for automatic identification, labeling and measurement, and (c) topological quantification of complex forms. The specific application given here was in the investigation of the branching pattern of the acinar pathways. The branching pattern was presented in a 3-D view and plotted in a 2-D fashion. Measuring the diameter and areas of the ducts along the center lines gave a summed curve of cross-sections. This reveals a Gaussian distribution with a maximum at about half a diameter from the origin. Numerous investigations show that alveolar ducts are completely ensheathed with alveoli as well as alveolar sacs as the blind-ending terminations of alveolar ducts.

The Gaussian distribution, unlike common belief, indicates an asymmetric branching pattern within the acinus but not ducts of equal length and symmetric branching. In a way, asymmetry of the respiratory units resembles the asymmetry in the conductive part of the bronchial tree of rat and mouse, as studied in the second part of this work. The knowledge of this distribution is the basis to model the functional issues of respiratory units, also being subject of the second part.

In humans, the size of the alveoli was reported to increase with higher branching generation (Haefeli-Bleuer and Weibel 1988), they frequently tend to form clusters with a common opening to the duct lumen (Hansen 1975). The number of alveoli per acinus was estimated to 2000 for the human lung based on global stereological estimations (Weibel 1963); other estimations give 494,000,000 alveoli for human lung, 19,700,000 for rat and 4,200,000 for mouse (Mercer 1988; Rodriguez 1987; Yeh 1979). For the acinus under study, a significant variation in the size of alveoli could not be found. With the help of the mean size of alveoli measured, i.e. 6.3×10^5 mm^3, one can conclude that approximately 430 alveoli are present in this acinus.

Fractal analysis of area occupation for bronchi and ducts/alveoli shows a remarkably higher percentage of area occupied by ducts/alveoli, but the increase in area measured towards higher resolutions has a lower slope. The reason for this might be in the function of the lung structure that is to bring fresh air without too much workload and resistance close to the acini but then reduce the speed of air, allowing diffusion into the ducts and respiratory exchange (Weibel 1993). The images and results suggest that the lung has a fractal structure to be described best by a Sierpinski carpet (Kaye 1989).

Scientific computing applied to 3-D microscopy gives rise to accurate and reliable concepts in computing and graphics to precisely interpret microstructures. Therefore, visualization and analysis are not mutually exclusive. This is quite obvious in the term "analytical graphics." In particular, any processing step to extract and identify volumetric substructures prior to quantitative analysis may be controlled by a 3-D reconstructed view. In addition, a number of visualization applications give evidence of integrated analytical tools to render specific volumetric subtleties. In order to fulfill such analytical tasks, visualization benefits from proven sets in digital image processing to enhance, filter, segment and code pictorial information, as required here to analyze the content of the acinar volume.

Various methods for 3-D visualization were discussed and tested. A specific volume rendering method using a back-to-front accumulation of voxel intensities and the flexible coloring and transparency setting worked best for these data. The problem of the complex and dense structural organization of lung was handled by extensive preprocessing and segmentation, combined with the use of color and a fine tuning of the opacity/transparency parameters. Various structural components can be enhanced. In addition, artificial lightning makes the perception of structures better still, in particular the circular alveoli. Invoking embedded geometry features made it possible to import the topological lines, giving analytical views. It seems that, in particular, this combination with volume rendering gives good insight into the structural organization of complex data. Also stereo displays and head-mounted displays were tested for the visualization. However, biological structures have free forms and generate far more surface elements than usual virtual reality applications, so that applications of these techniques are still limited to very simple forms.

The time required to investigate acini compared to a wax plate technique can be reduced by a least a factor of 10 by the techniques presented. The results of this study encourage the application of this method to a wider set of samples.

6 Analysis of the Conductive Part of Lung

6.1
Introduction

Up to now it was technically difficult to represent a complete bronchial tree of lung with all structural details in the form of a computer model. However, with the progress in 3-D image acquisition, processing, visualization and computer modeling a complete representation of mammalian lungs has come within reach. The use of casts is perhaps still the most effective way to study the macroscopic branching pattern of smaller species, and the 3-D representation and measurement based on stereoscopic representations is used here. For forms larger in space, such as the human lung, recent technological progress in high resolution computer tomographic imaging (HRCT) has made 3-D imaging of the bronchial tree possible.

The computer modeling discussed here concerns the structure of the rat lung, which has about 4000–5000 bronchial segments. Each of these tubes, either bronchi or bronchioli, always has two daughter segments. Casts of a half-adult rat will be investigated which match the developmental stage of the previously studied histological material (see the first part of this work). The approach suggested here is to first measure the main branches of a lung exactly and analyze selected parts of the bronchial tree in detail and, secondly, to use obvious regularities and similarities present in the lung structure to complete a computer model by means of a fractal graphics. Stochastic restrictionby using the data of the main branches and matching against measured data improves the degree of proximity. As a result, a computer model of a complete bronchial tree will be designed to study physiology and function after additional implementation of the necessary physical issues.

Specific results of the following chapters include the development of methods to: (a) generate a statistical, topological description of 3-D branching patterns out of stereoscopic tracings of casts, (b) evaluate a realistic model of lung morphology by combining measured data with fractal graphics, and (c) apply computational physics to achieve a functional model.

Figure 33 gives an overview of the programs developed and the exchange and flow of data. The following text describes these programs, their function and application.

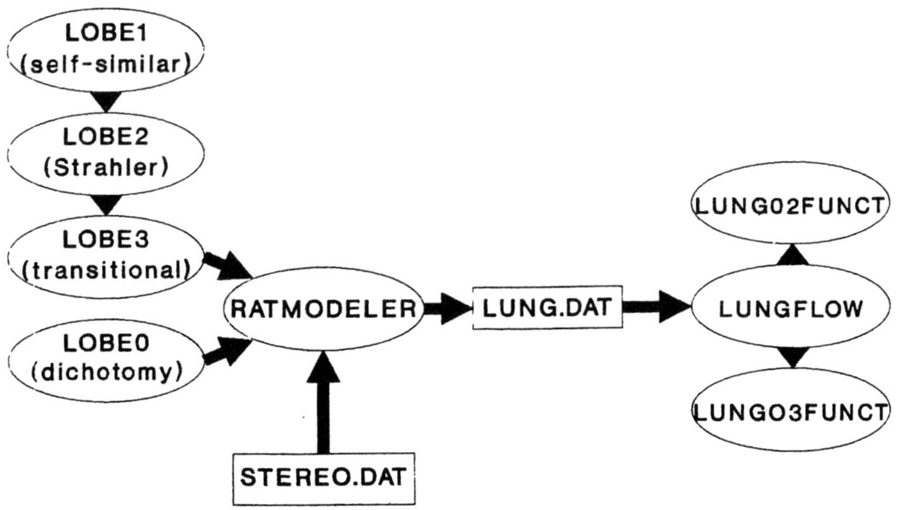

Fig. 33. Dataflow and programs developed to model morphology and function of the bronchial tree of rat

6.2
Stereoscopic Tracings of Casts

Plastic casts were made available from K.-P-Valerius. As described in his thesis (K.-P.Valerius 1992), a silicon rubber material was used at 20 cmH$_2$O (1 cmH$_2$O=98.1 Pa, 20 cmH$_2$O=1962 Pa) after freeze fixation. One example of such a complete lung is given in Fig. 34. From this particular cast, the major branches were analyzed. In addition, a number of individual lobes were investigated as well, such as the diaphragmatic lobe and the apical lobe displayed in Figs. 5 and 36. Usually five types of lung lobes (right apical lobe, right cardiac lobe, right intermediate lobe, right diaphragmatic lobe, left lung lobe) are differentiated. Although the left lung lobe is not divided up further and more simply constructed, corresponding segments to the right apical, intermediate and diaphragmatic lobes can be identified, which is a helpful identification for a data representation.

Stereo images were taken with a macro-lens of corrosion cast models under well-adjusted illumination. The blue looking casts were additionally sputtered with a layer of thin gold to enhance the contrast and visibility. The stereo angle was 6°–9°, and resolution of the images after digitization was 512×512 pixels.

The Sterecon system, developed at the New York State Department of Health in Albany was used to trace the branches (Marko and Leith 1992). The system uses a two-monitor display of the digital images, arranged 45° from a half-silvered mirror (Fig. 37). Circular-polarizing screens of opposite direction were mounted in front of

Fig. 34. a Corrosion cast of a rat lung, 25 mm in length. **b** Corresponding 2-D projection of traced centerlines. *Numbers* have been assigned to the bronchial segments; diameters in mm are given in *parentheses*

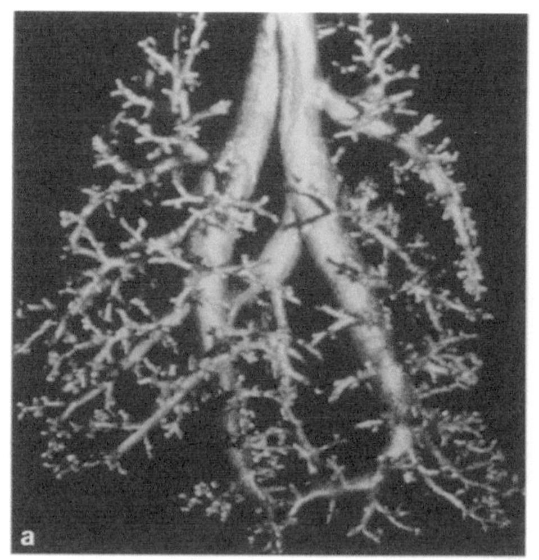

a

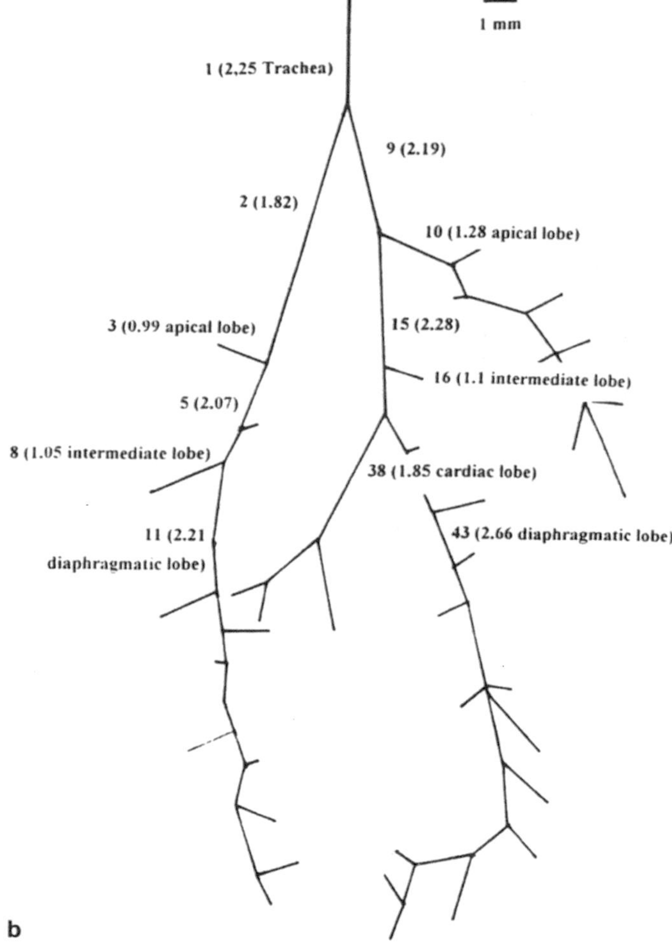

1 mm
1 (2,25 Trachea)
9 (2.19)
2 (1.82)
10 (1.28 apical lobe)
3 (0.99 apical lobe)
15 (2.28)
16 (1.1 intermediate lobe)
5 (2.07)
8 (1.05 intermediate lobe)
38 (1.85 cardiac lobe)
11 (2.21 diaphragmatic lobe)
43 (2.66 diaphragmatic lobe)
b

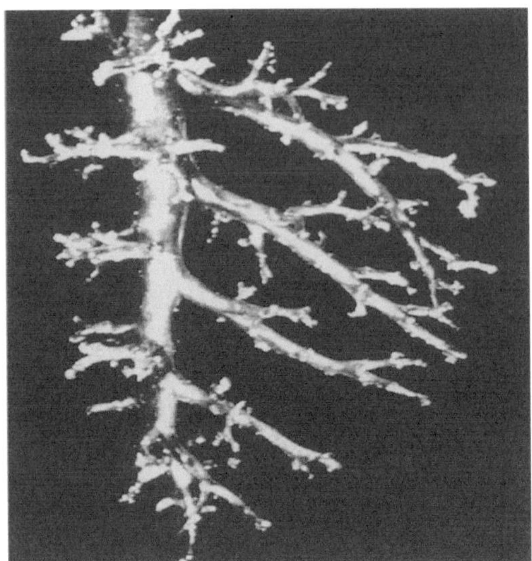

Fig. 35. Cast of diaphragmatic lobe

Fig. 36. Cast of apical lobe

each monitor. The mirror fuses the two monitor images into a stereo–image, resolved for the observer by wearing polarizing glasses whose polarizing direction corresponds to the filters of the monitor. The begin and starting points of the lung segments were indicated and lines were drawn in stereo overlay bit planes on the monitors with a digitizing tablet. Moving the digitizer controls the position of a 3-D cursor on the screen (in x/y), whilst additional knobs on the digitizer control for depth position of the cursor (z). The x/y/z coordinates of the begin and start end points of segments were stored in an ASCII-file format.

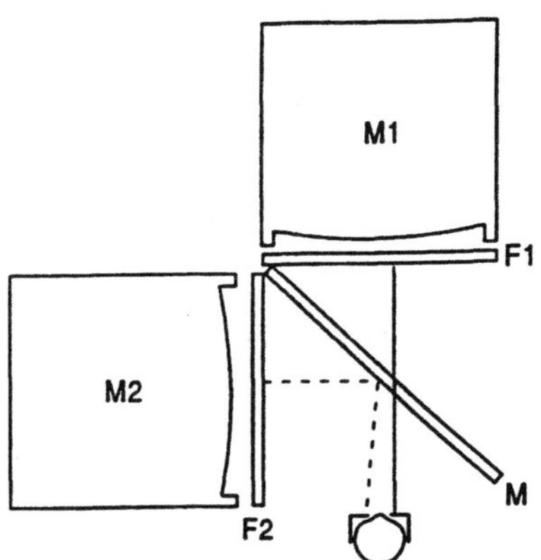

Fig. 37. Stereo-display of the STERECON-System. Two monitors (*M1,M2*) with polarizing filters (*F1,F2*) display a stereo pair, fused by a half-silvered mirror *M*. The observer wears polarizing glases (from Marko and Leith, 1992)

6.3
Analysis of Traced Data

A two-dimensional projection of the main branches of a complete lung are given in Fig. 34b. The assignment of the numbers for the individual segments is for internal use only, but mark the way of measurement performed. Numbers in brackets give the diameter of the segments in millimeters. With the same procedure, individual lobes were measured, including those depicted in Figs. 35 and 36.

These data are subjected to an analysis of the topology and its statistics. Typical parameters include: (a) individual length, diameter and branching angles; (b) sum of path length; (c) mean of path length, diameters and branch angles; and (d) frequency distribution of generation numbers, length and angles. These parameters provide the necessary database before a modeling can commence.

The apical lobe shown in Fig. 34 documents a dichotomous branching pattern. This corresponds to the public belief of how a lung structure looks. However, a typical lung lobe is much more characterized by its asymmetry, as depicted in Fig. 35. The branching pattern has a "strong" side and a "weak" side of the first order branches of the lobe. First order bronchial segments are defined here as those, which have their origin at the main stem bronchi, a straight arrangements of thicker and longer segments. The asymmetric form is essential for a monopodial branching pattern.

It appears that monopodial forms exist without asymmetry. This holds for the intermediate lobe, since in some areas of this lobe the left and right bronchioli are of same length and angle. If the reason for this transition is in the evolution of lung structure, it might be of importance. In any way, the development of the symmetric form appears more likely to be related with the asymmetric monopodial arrangement, whilst a change from a dichotomous branching towards a symmetric monopodial one is difficult to understand, even when an irregular dichotomy is taken into account.

Closer inspection reveals that for most lobes asymmetry is also present for the second and third order branches. In a way, the bronchial tree seems to be self-similar, at least within these three levels. For further approval, the length of the second order

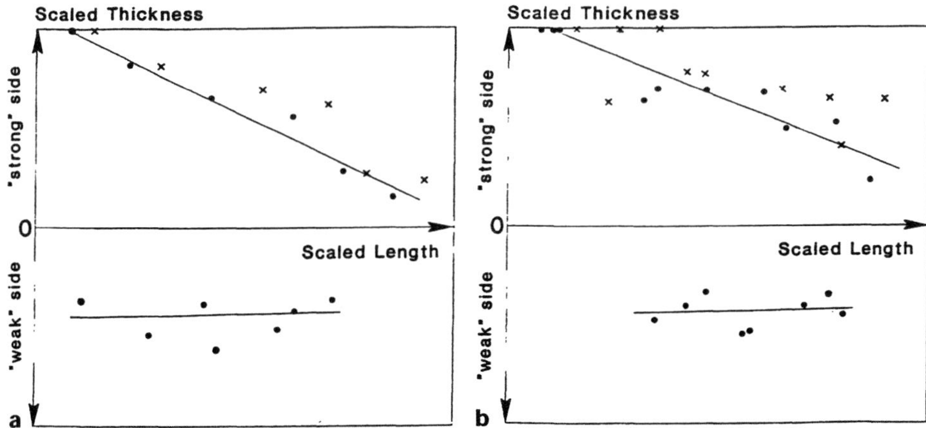

Fig. 38a,b. Self-similarity of assymetric lobes. Scaled thickness of first order branches (**a**) and of second- and third-order branches (**b**) of both sides of the lobe is referenced to the scaled length of these branches. The strong sides of two lobes have been measured (*dots and crosses*)

branches was scaled versus the diameter of the branches. This was done separately for both the "strong" and "weak" side of the diaphragmatic lobe (see Fig. 38a). Same procedure was carried out for the 3rd order branches. The result is given in Fig. 38b. Comparing the two drawings gives a similar slope in each. The "weak" sides have only limited gradients; these branches change their length relatively less than the "strong" side branches. In addition, at the periphery of the second and third order branches similar length/diameter relationships of the two sides are likely to occur. This may be interpreted as an increasing dichotomous branching pattern more distal.

7 A Computer Lung Modeler

7.1
Introduction

Modeling of the lungs branching pattern has been a subject of the mathematical treatment by fractal geometry, including Mandelbrot (1983). The essence of calculating such forms is in the use of recursive algorithms. For the description of plant structure and development Lindenmayer (1968) introduced L-systems. The L-system is used to simulate the growth of structures by reproduction with the help of fractal templates (Rozenberg and Salomaa 1992; McCormick and Mulchandani 1994).

The generation of objects is best described as a (software) command string rewriting, whereby complex objects are constructed out of a simple initial object. Fritjers and Lindenmayer (1974) were the first to introduce a graphical interpretation later this concept was extended to an interpretation by turtle geometry (Prusinkiewicz and Lindenmayer 1990). A virtual turtle is used to follow the outcome of the command strings (such as move forward, left, right). These concepts can be generally extended to three dimensions.

One issue of the fractal modeler presented here is in the application of Lindenmayer (L)-systems. A L-system will be adapted to this application. Essentially all recursiveness and self-similarity is coded in a C program, which allows a reproducible tuning to match against available topological databases.

7.2
A Self-Similar, Asymmetric Model
of a Lung Lobe

A first and simple way to model the bronchial tree is a regular, dichotomous branching pattern. Such a model has become very popular to assign the structure of lung (Kaye 1989; Oliver and Hovis 1994). A outcome of such program is given in Fig. 39a (program LOBE0). Such pattern approximate the structure of an apical lung lobe (see Fig. 35). However, it is quite obvious, with reference to the lobe given in Fig. 36 that such a model does not represent these casts correctly. It also turned out to be very difficult to transfer the program generating dichotomy into a generator of monopodial arrangements. Therefore, another series of programs was developed.

With respect to the asymmetry in that there are daughter branches of different size, pointing into opposite directions, the first outcome of the modeler is depicted in Fig. 39b (program LOBE1). A high degree of self-similarity is used, since the pattern

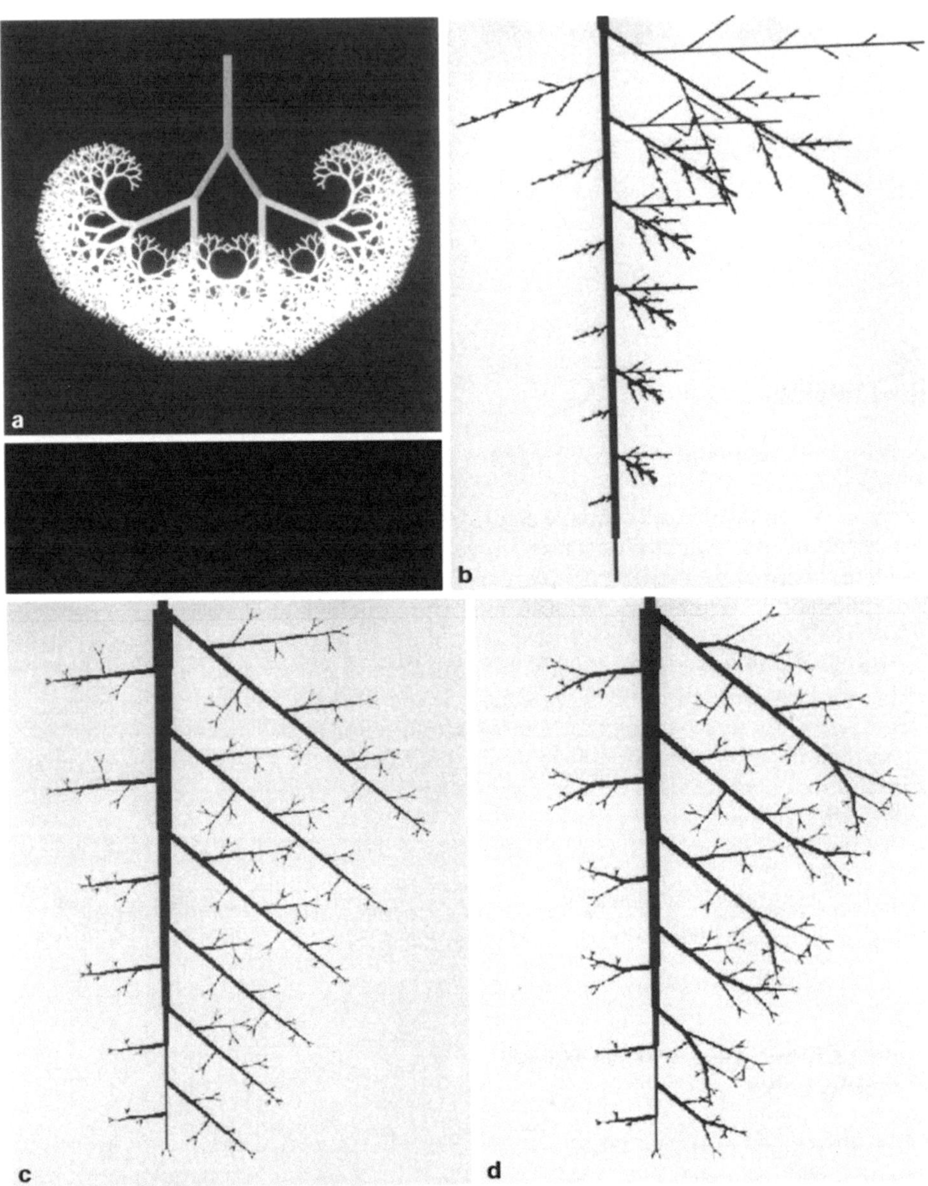

Fig. 39a–d. Development of a computer lung modeler based on a fractal Lindenmayer-system. **a** Dichotomic scaling; **b** assymetric, self-similar model; **c** introduction of scaling and Strahler – ordering scheme; **d** transition to a distal dichotomic branching

of the first order is used as a template for the next order branches. All branches are scaled according to their branching order.

7.3
Scaling and Strahler-Ordering Scheme

The model presented above is limited by the high degree of self-similarity which includes the scaling. This however has little value for lung lobes, since such a model would imply a decreasing scaling in the distal segments originating in the main stem as having the same complexity as branches originating more proximally. The overall self-similarity must be limited and the degree of order must be included, in the sense that segments of identical order have identical length and that the order of branches originating distally and on the "weak" side is less.

Taking the Strahler ordering scheme, a branch of a certain order has daughter branches of the next higher order, but also daughter branches with more than just one higher order to assure asymmetry. The program LOBE1 does not include variations in the branching angles either, but there is a close relation between (order) stepsize and the branching angle (Horsfield 1983). This peculiarity can be documented in the topological data sets comparing segment length, angle and diameter. With this an improved model is given by the modeler which is shown in Fig. 39c (program LOBE2).

7.4
Transition in the Bifurcation Pattern

Comparing Figs. 39c and 35, still shows differences can still be seen in the more distal parts of the bronchial tree arrangements. Several authors state that the alveoli of the mammalian lung have about the same size. Also, the diameter of the terminal bronchioles was found to be correlated with the volume of acini (Rodriguez et al. 1987). Moreover, there is a correlation between the branching angle of bronchi and bronchioli and the distal volume they supply (Horsfield 1985). One can therefore conclude that the branching pattern of the conductive part of lung ends in a more symmetric, dichotomous form. In addition to functional reasons, this may be partly because the bronchial segments are oriented perpendicular to the outer lung boundary. For equal filling and dense packing of acini other than asymmetric branchings must be present; regular deviations from this may give rise to shell-like arrangements below the surface of lung. Since it is also obvious from inspection of the topology of cast models that the first branching orders are monopodially arranged, a transition from a monopodial to a dichotomous branching pattern is suggested. Modeling the form of the complete conductive part of lung must incorporate these facts. Branching models available so far up did not consider this transition. Figure 4c gives an example of such a modified model (program LOBE3). The form of this lobe very much resembles the original form (Fig. 35).

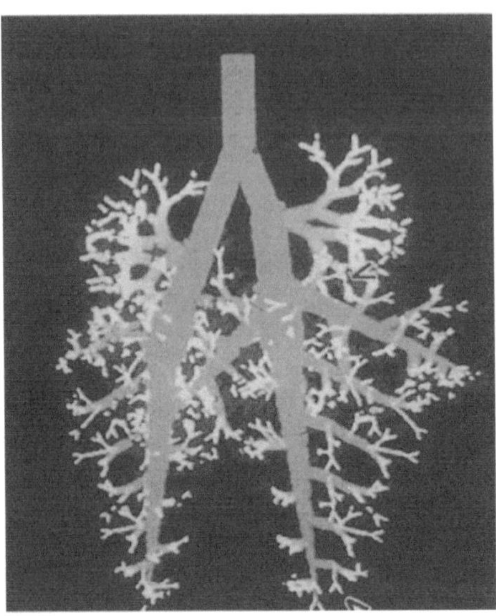

Fig. 40. Complete morphometrical model of a rat lung. Sites near the stem of the intermediate lobe and at the diaphragmatic lobe are marked to be used for "digital" dosimetry

7.5
Completing a Graphical Lung Model with Limited Stochastics

The final step in modeling a complete organ is in the combination of measured properties and arrangements of lobes originating in the main stems and their 3-D arrangement. Based on the lobe modeler described above, measured 3-D spatial data of the main stems are now combined with fractal graphics. The size of the lobes is governed by the diameter of the main first bronchial segment. The program LOBE0 (dichotomy) was carried out for the modeling of the apical lobe and its counterpart of the left pulmonary lobe LOBE3 (asymmetry) assisted in modeling the intermediate, diaphragmatic lobes and their counterparts of the left lung lobe and the cardiac lobe. Adaptation has been included for correct size, left–right symmetry and spatial orientation of the lobes in three dimensions. Figure 40 depicts the completed conductive model of a rat lung, demonstrating a template of a cast given in Fig. 33 (program Lungmodeler).

The program generates a bronchial tree featuring 4090 bronchial segments down to terminating bronchioli and since on terminal bronchioli supplies to one acini 1999 acini are present. Segments are defined as the individual tubes of the bronchial tree limited by bifurcations. These bronchial segments must be differentiated further from the lung segments, which are larger subunits of the lung lobes.

This model is based on a half-adult rat (*Rattus norvegicus*), having a weight of 237 g (Valerius 1992). The number of bronchial segments is slightly less than what is given by the model of Yeh et al. (1979), with 6123 tubes and 2487 acini for a adult rat of weight 330 g. For the computer model, the smaller left lobe has 1574 bronchial segments, the right lobe 2516 segments (1:1.6). Yeh's model gives 2048 for the left and 4075 bronchial segments for the right lobe (1:1.98). The individual lobes of the computer lung have the following number of bronchial segments: 511 for the right apical lobe, 560 for the right cardiac, 667 for the right intermediate, and 774 for the right diaphragmatic lobe.

All the spatial coordinates of the beginning and end of the bronchial segments are stored in an ASCII file (LUNG.DAT). Together with the coordinates, for every segment the running number, model length, model diameter, the running number of the corresponding daughter and parent segments are stored. In a separate step, the distal amount of volume supplied was calculated for each segment. This is done by summing up the volumes of all tubes, beginning at the terminating bronchioli and proceeding proximal up to the trachea.

8 Computational Physics Applied to a Bronchial Tree Model

8.1
Introduction

Based on a computer lung exhibiting a model of the bronchial tree the gas distribution and take-up is simulated by solving mass transport equations. Since functional characteristics of a particular segment influence the input to the daughter segments, only a top-down computational solution is possible, starting from the trachea and proceeding distally. Since all bronchial segments are handled individually, the calculation is a type of finite elements. To give a stable solution, for each of the segments parameters such as flow, diffusion and uptake must be calculated for a very short time span of the inhalation or expiration interval to obtain stable numerics. By using 3-D computer graphics and color coding, the flow or gas concentration can be displayed on the computer screen to give insight into function under various breathing conditions.

8.2
Scaling of the Computer Lung Model

So far, the computer model, when displayed with computer graphic tools, shows a particular structure, which is correct for the relative length of bronchial segments and branching angles.

The thickness for graphical display is used in a way that the minimal thickness, in particular for the smallest branches (terminating bronchioli), is 1 pixel in thickness. What is required is to internally scale this graphical model into a realistic, physical model matching real dimensions in length, thickness and volumes and include the dynamics during breathing. This is also a prerequisite for functional calculations and predictions.

Diameters of the main stem bronchi have been taken from measurements (see Fig. 34). An idealized model based on cast measurements has been published by Yeh and coworkers (1979). The advantage of his findings lies in the differentiated investigation for individual lobes. The diameter versus length is graphically displayed in Fig. 41, related to the branching order. All lobes exhibit similar gradients in bronchial size; basically, the "offset" varies. The results have been taken to scale the computer model with these findings, from the first segment where the lobe originates down to the terminating bronchioli, which have a length of 0.35 mm and a diameter of 0.2 mm.

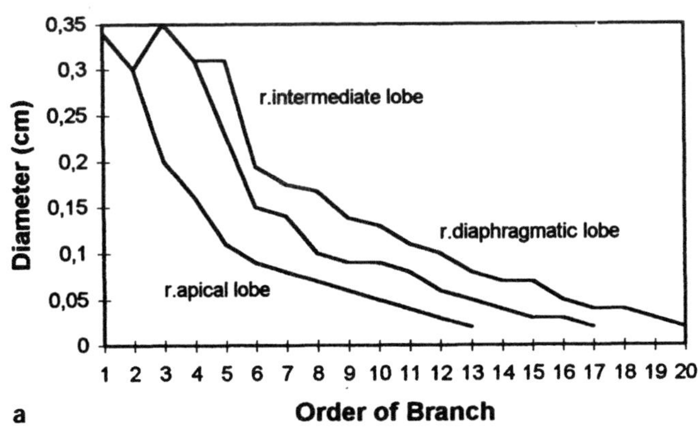

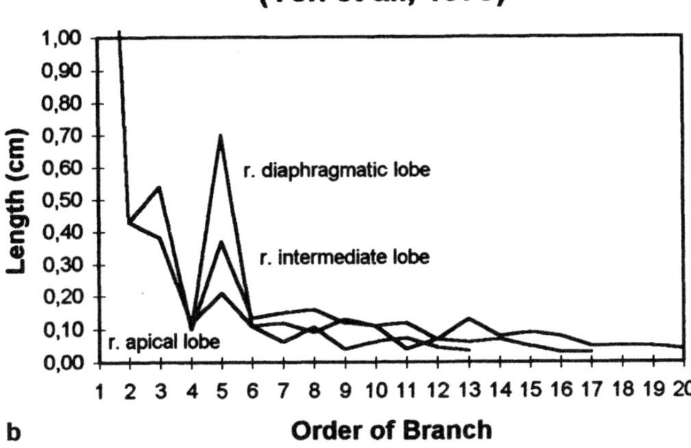

Fig. 41. **a** Diameter of bronchial segments of an idealized model of Yeh. **b** Corresponding lengths. These findings are used to internally scale the computer model

From this scaling, the total volume for the conductive part of the lung model is 0.69 ml. To compare this value with other findings, the pressures applied during preparation are of importance. Values of Valerius (1992) for the conductive part of the bronchial tree of a rat (*Rattus norvegicus*, 237 g, 20 cmH$_2$O) are 0.8 ml, being 5.6% of a total volume of 14.4 ml. The corresponding volume of the acini is 13.6 ml, respectively. Yeh reports a 9.2-ml acinar volume at 13.7 total lung capacity for a pressure at 10 cmH$_2$O. In this finding, the conductive part of the bronchial tree makes 8% of the total volume. Other authors report similar values, such as 0.77 ml for the conductive and 8.75 ml for the respiratory volume (Rodriguez et al. 1987).

For this computer model, the total acinar volume was set to 7.00 ml at a model pressure of 20 cmH$_2$O, i.e., 10% of the total volume is in the conductive part.

8.3
Dynamics with Breathing

8.3.1
Model Pressure and Volumes

There is a functional relation between lung inflation, gas flow and transpulmonary pressure (Hyatt 1963). Pressure gradients are not considered directly in this model, except for a calculation of resistance, for the following reasons: (a) For normal breathing it was shown that the frequency-dependent behavior of the lung is not significant (Otis 1956). (b) The pressure gradient between trachea and alveoli is assumed to be zero for slow breathing or for normal breathing (Ross 1957). The flow rate of every bronchial segment is proportional to the distal volume supplied; flow partitioning at bifurcations considers this fact. (c) Related to the above is the assumption that there are no pressure gradients between individual alveoli or groups of alveoli, because it was shown that the surface-to-volume relation is constant, being a measure of the degree of expansion (Mercer 1991). This assumption also excludes asynchronous ventilation and high-frequency flows (*Pendelluft*; Slutsky et al. 1985).

For the computer model, the term pressure is used only to assign a certain volume inspired and to link this volume to the size of alveoli and their characteristics such as size, surface and ratio of ducts to alveoli volume. This scaling in the computer model is done in a way that a pressure of 0 cmH$_2$O is assumed to describe the lung structure at residual volume and a pressure of 30 cmH$_2$O is at maximum inspiration. Functional residual volume (FRV) in this model is at 5 cmH$_2$O, for normal breathing the region above 5 cmH$_2$O up to 15 cmH$_2$O is used, breathing at maximum vital capacity (MVC) is from 0 to 30 cmH$_2$O. Vital capacity, together with the residual volume, gives the maximum lung capacity. A hysteresis for differences in the pressure/volume relationship between inspiration and expiration is not included in the model.

An initial value of 7.69 ml for the complete gas volume of the model matches the gas volume at the deflation curve found by Mercer et al. (1987). Since essentially only the volume of the acini increases with inspiration and the conductive volume remains unchanged, volumetric values for 0, 5, 10 and 30 cmH$_2$O pressure of the total gas volume were related to these measurements and all other values can be calculated from here.

Typical model pressure/volume relationships including total volume, volume increase (dV) and the mean volume and radius of the respiratory units are presented in Table 1. The residual volume of the computer model is 1.3 ml. Normal breathing is above functional residual capacity of 3.2 ml. Maximum lung capacity is 8.5 ml. The increase in volume during inspiration is linear, but slightly saturated towards maximal capacity.

8.3.2
Modeling of Respiratory Units and Volume-to-Surface Relationships

Three types of acini have been defined for the computer model, to study variances in functional behavior (Fig. 42):
- Type 1: small; frequency is 25% of total population; radius is 0.95 mm; volume is 2.5 mm^3
- Type 2: medium; frequency is 50% of total population; radius is 1.195 mm; volume is 3.5 mm^3
- Type 3: large; frequency is 25% of total population; radius is 1.365 mm; volume is 4.5 mm^3

Size and volumes hold for a model pressure of 20 cmH$_2$O. The rationale behind using different types of acini is in the consideration of variances of sizes found by various authors (e.g., Mercer et al. 1991; Rodriguez et al. 1987) and the possibility to study differences in function.

The corresponding values of Valerius (1992)/ at 20 cmH$_2$O is a mean acinar volume of 10.27 mm^3; this gives 1324 acini only (variation coefficient 5). Mercer et al. report a measured volume of 0.53 mm^3 based on serial sectioning (Mercer et al. 1991). Rodriguez et al. (1987) report a size of 1.86 mm^3 at 60% of lung capacity. The great variances may be caused by the preparation procedures, since studies using casts give values up to three times higher than histological studies. Shrinkage in the embedding, cutting and staining procedures may also have lead to an underestimation of acinar volumes and compartments in the first part of this study. This, however, does not have an effect on the relative structural relationships.

For functional computations, in the sense of finite elements, the volumes of the acini are assumed to be composed of ten individual volumetric segments or subsections. Figure 43 gives the number of alveoli present in these acinar segments of all three types of modeled acini.

The volume increase in these segments of a type 2 acini is displayed in Fig. 44, at model pressures of 0 and 30 cmH$_2$O (range of high tidal volumes) and 5–15 cmH$_2$O at normal breathing.

For the determination of surface-to-volume relationships, essential for the prediction of gas take-up, the fluctuations of the volumetric structural composition in

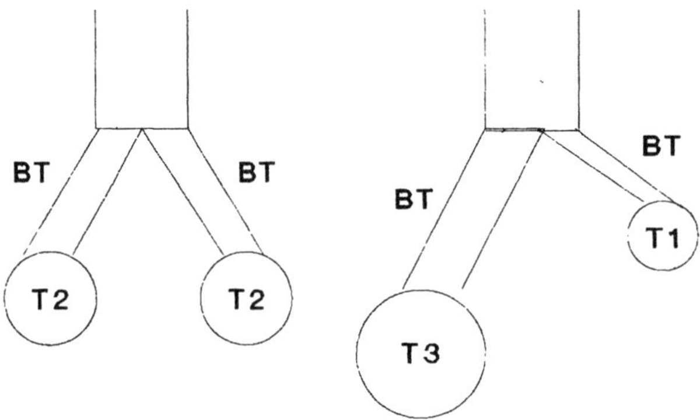

Fig. 42. Arrangement of three types of acini (T) within the computer model

Distribution of Alveoli in modeled Acini

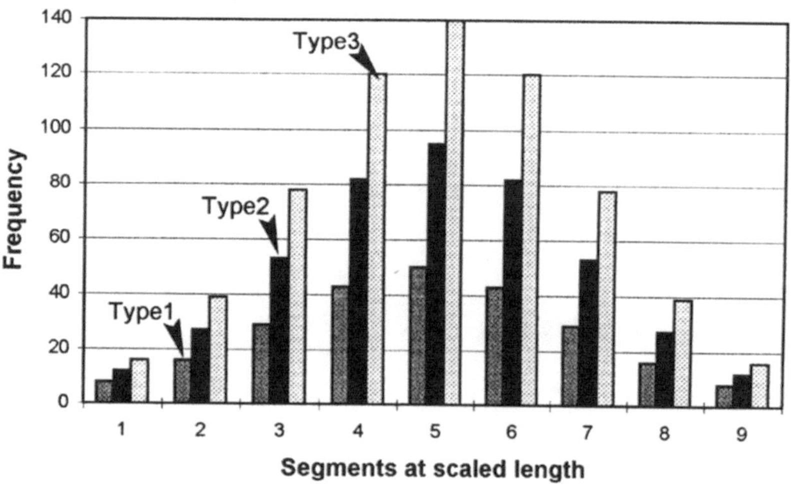

Fig. 43. Frequency of alveloi in three types of modeled acini, divided into nine segments

Acinar Volumes at various Model Pressures

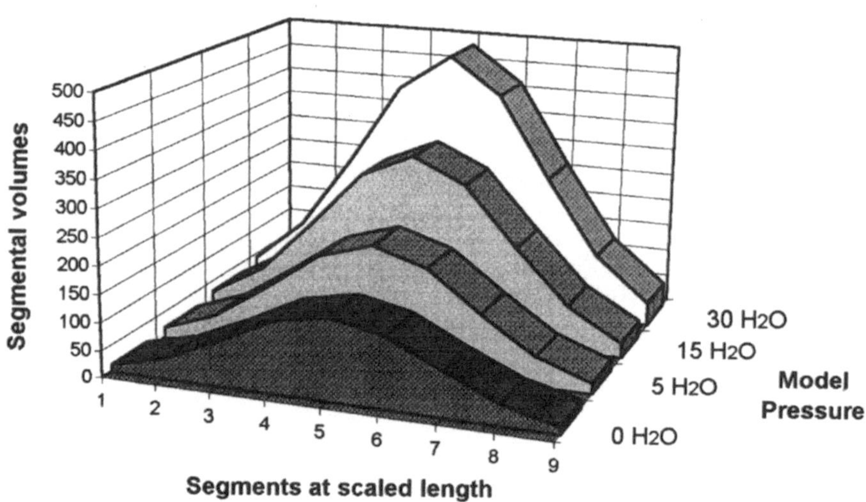

Fig. 44. Increase of volume of a type 2 acini at different model pressures

the respiratory units must be known. The elastic and surface tension properties of the alveoli are predominant in the volume range around functional residual capacity and the percentage of volume in the alveoli is larger than in the ducts. The elastic network, particularly around the alveolar ducts, determines the behavior near full volume expansion, and the percentage of volume is almost 40% ducts volume to 60% alveolar volume above 10 cmH$_2$O pressure (Mercer et al. 1987). This matches well the relation of volumes of the acini studied in Part I of this work at a specimen preparation of 20 cmH$_2$O (relation of ducts to alveolar volume is 1:1.46).

Since the surface is in direct relation to the volume, a surface-to-volume ratio of 37.5 was used.

8.3.3
Flow Rates and Initial Partial Pressure

Inspiration flow rate was set to 8 ml/s, 80 breaths/min, with equal inspiratory and expiratory times of 0.3 s. As an example for normal breathing, 2.4 ml are inspired in 0.3 s. For higher tidal volumes the time of one breathing cycle was not changed (Mercer et al. 1991).

Initially, the concentration in all segments was set to oxygen backpressure of 5.26%. With the begin of breathing, the first segment gains oxygen from the outside with a concentration which was set to 21% for fresh air. No diffusion was assumed for this first segment, since the speed of air is very high in the trachea.

8.4
Convection

The sum of the cross-sectional area of bronchial daughter segments (A_{d1}, A_{d2}) is usually less than the area of the parent airway (A_p), and the two cross-sectional areas of the daughter airways are also different. Equations describing flow partitioning take into account not only the variance in cross-sectional areas, but also the distal ventilatory volumes V_{tot} of both daughter segments and the individual distal volume V_{d1} and V_{d2}. Consequently, the flow partitioning factor must be calculated first for all segments, based on the distal volumes already present in the data set (LUNG.DAT).

For a particular daughter segment, the flow v_{d1} is (Eq. 6):

$$v_{d1} = (v_p \, A_p \, V_{d2})/(A_{d1} \, V_{tot}) \tag{6}$$

with v_p being the flow in the parent segment. Behind this equation is the assumption that the distal volumes are filled equally and to prevent an unbalanced flow or the so called *Pendelluft* phenomena. It was found to be a reasonable assumption that the flow rate in any bronchial segment is related to the number of terminal bronchi evolving from it, and not from its cross-section only (Ross 1957).

8.5
Resistance and Reynolds Number

Resistance is present for any moving part of the respiratory system. This includes airflow resistance in the tracheobronchial tree as well as pulmonary-tissue resistance, also summarized in the term "pulmonary resistance." If chest-wall resistance is also included, this is named as "thoracic resistance" (Dubois 1964). For air-flow resistance one must differentiate laminar and turbulent flow. Flow will be turbulent, when the inertial force being the source of turbulent energy overcomes the viscous friction force. The latter tends to dissipate turbulent energy. The Reynolds number expresses this relationship between the two forces.

The Reynolds number Re in aerodynamics can be calculated in various ways. For the bronchial tree composed of individual tubes the Re number is a function of length, speed of flow v, density d and dynamic viscosity. The density of air is 1.2928 kg/m^3 at 0°C. The relation of density and dynamic viscosity can be combined into a kinematic viscosity η, which is 0.00001438 (m^2/s) at 15°C and a pressure of 760 torr. For tubes the diameter d is taken instead of length. Then, Re is (Eq. 7):

$$Re = v * d/\eta \qquad (7)$$

Flow is assumed to be laminar at values of Re and turbulent at Re 4000. Since in the bronchial tree the flow and the diameter of the segments decrease more distal, the highest Re number is to be expected in the trachea (Ultman 1983).

To maintain a streamlined flow through a tube, a certain pressure gradient is necessary, depending on the volume rate of flow V and the geometry of the tube such as length, cross-section A and the kinematic viscosity of air (Poiseuille's law; Eq. 8):

$$R = dp/dV = 8 \; \pi \eta l/A^{2.} \qquad (8)$$

This relation is also known as Poiseuille's law (Ross 1975).

8.6
Diffusion

Diffusion is a transport phenomena, which tends to diminish nonuniform distribution within a gas or a gas mixture. The nonuniformity within a volume of interest may be introduced by regional pressure, thermal energy or different kinds of molecules.

Due to a constant thermal motion of molecules in their gaseous state, these molecules more likely move in the direction in which the density (the numbers of molecules of a given kind) is less. Diffusion tends to increase the entropy of the distribution of molecules in a given volume.

Supposing a volume having a concentration varying in one direction, assumable x, then the number of molecules per unit volume or the concentration C is a function of the x-coordinate alone that is C(x) expressed in m^{-3}. Diffusion then occurs in the direction of x. The partial current density, j, is the net number of molecules crossing a unit area placed perpendicular to the direction of diffusion and is expressed in m^{-2} s^{-1}. This mass transport is proportional to the change in concentration per unit length or the concentration gradient dC/dx. Most diffusion processes therefore satisfy the relationship (Eq. 9):

$$j=D \ (dC/dx) \tag{9}$$

where D is a proportionality constant called diffusion coefficient which depends on the substance and is expressed in square meters. D in particular depends on the self diffusion of the molecules and is in linear relation to the mean free path and the velocity of the molecules (Chang 1985). This equation was suggested by the German physiologist Fick (1829–1901).

The above equation does not use the time variable explicitly. In order to describe unsteady diffusion, a variation of the wave equation can be used (Alonso-Finn 1967). The wave equation is fundamental in physics to describe the evolution in time and the distribution in space of an undisturbed and not attenuated field. For other phenomena, such as the transport or net transfer of matter, energy or momentum in a bulk or macroscopic amount, a propagation equation can be defined as (Eq 10):

$$dF/dt=a^2 \ d^2f/dx^2 \tag{10}$$

with F as the field corresponding to a particular transport phenomena and a describes a constant characteristic. For diffusion, the equation for the 1-dimensional case is as follows (Eq. 11):

$$dC/dt=D \ d^2C/dx^2 \tag{11}$$

This equation shows the dependence of C (the concentration) from the second derivative with respect to space and the first derivative with respect to space. Implicitly, the principle of conservation of mass is considered, stating that the rate of accumulation of a given substance in the infinitesimal segment of interest is equal to the rate of mass brought into and removed from that segment.

In steady state, the number of particles entering and leaving a volume element always remains constant and no accumulation occurs (i.e., j=const). For a rod or pipe of length l, and concentrations C0 and C1 kept constant on the two sides of this pipe, this gives (Eq. 12, Alonso-Finn 1967):

$$j=D \ (C0–C1)/l \tag{12}$$

8.6.1
Equations Describing Diffusion in Lung

Of importance for lung simulations is the non-steady-state. At a given time a higher concentration enters the pipe (bronchiole), and depending on whether the pipe is open (bronchiole) or closed (alveolar sac), various concentration distributions may occur. The mass transport equation describing this situation, if we add the area a being the cross-sectional area of the pipe, is (Eq. 13):

$$j0-j1/dt=D \ a \ dC0-dC1/l \tag{13}$$

As one notices, this equation, in accordance with unsteady diffusion, depends on time and on the difference in the concentration gradients dC0 and dC1 on both ends of the pipe. Solving the diffusion equation for such a case is a difficult problem in mathematical physics, since no direct analytical approach for these kind of differential equations exists. One way of solving this differential equation is in a numerical, iterative way.

The situation in lung is slightly more complicated, since a segment of interest stems upstream from a branch and is divided downstream into two branches. For the mass transport into the segment, the dispersion at the branch must be known. In addition, the concentration cb1 and cb2 in the two daughter segments with the cross-sectional areas ab1 and ab2 must be considered for the mass transport out of the segment.

The effective dispersion coefficient is described by the total Vt and distal volumes Vd, together with the molecular diffusion coefficient Dm and the length of the branches (Eq. 14, Mercer et al. 1991):

$$D_e = D_m - \gamma \, l \, \frac{v}{3} \qquad \text{with} \qquad \begin{aligned} &\gamma = 1.08, \, v > 0 \text{ (inspiration)} \\ &\gamma = 0.37, \, v < 0 \text{ (expiration)} \end{aligned} \tag{14}$$

The dispersion for a particular branch is (Eq. 15):

$$D_j = \frac{l_{j-i}}{l_{j-1} + l_j} \, D_{ej} + \frac{l_j}{l_{j-1} + l_j} \, D_{e\,j-1} \tag{15}$$

8.7
Mass Transport Equations

The morphological computer model can be extended to a functional one by the dimensions given for all segments and with the knowledge of distal paths and distal ventilatory unit volumes. For this purpose, mass transport due to *convection, diffusion* and *mass uptake* must be calculated. This is done in an iterative way for all bronchial segments and acini forming a finite element construction. To these finite elements mass is added or removed by convection, diffusion or uptake. The mass transport is of type non-steady-state due to inhalation and expiration. Such an approach was described previously for reactive gas dosimetry (Miller et al. 1978; Mercer et al. 1991).

The corresponding equation describing the mass change is:

$$\frac{d_{mass}}{d_{ume}} = \text{diffusion} - \text{convection} - \text{uptake} \tag{16}$$

The change within a segment j and time steps i and i+1 can be represented more explicitly by using:

$$\frac{m_{j,\,i+1} - m_{j,\,i}}{dt} = D_j a_j \, \frac{\Delta_D C}{l_j} - v_j a_j \, \Delta_u C - K_{aw} \, S_v \, a_j \, L_j \, C_j \tag{17}$$

where $\Delta_u C$ is the difference in concentration across segments, $\Delta_D C$ is the difference in contration gradient, K_{aw} is the mass transport coefficient, a the cross-sectional area of the segment and S_v is the surface-to-volume ratio. The change in concentration gradient in one element j per time step dt can then be calculated using concentrations numbers C in the daughter elements s1 and s2:

$$\frac{C_{j,i-1}\, A_j\, l_j - C_{j,i}\, A_j\, l_j}{dt} = D_{j,i}\, a_j \frac{C_{j-1} - C_j}{(l_{j-1} + l_j)/2} - D_{s1}\, a_{s1} \frac{C_{j-1} - C_j}{(l_{j-1} + l_j)/2} - D_{s2}\, a_{s2} \frac{C_{j-1} - C_j}{(l_{j-1} + l_j)/2}$$

$$+ C_{j-1}\, a_j\, v_j - C_j\, (a_{s1}\, v_{s1} + a_{s2}\, v_2) - K_{aw}\, S_{vj}\, A_j\, l_j\, C_j \tag{18}$$

with $A_j = {}^1/_3\, [a_j + a_{s1} + a_{s2} + \sqrt{a_j\,(a_{s1} + a_{s2})}]$.

The first term uses the difference in concentration gradient of a segment for diffusion, since the second term uses the difference of concentration for convection; K_{aw} is the mass transport coefficient and S_v the surface-to-volume ratio in the term describing mass uptake. Aj is a weighted number for considering the cross-sectional areas to be "tapered," instead of using a constant-radius airway segment. This takes into account the fact that the cross-sectional areas of the daughter segments do not exactly equal the sum of the daughter airway segment. Similar assumptions have been included for ozone dosimetry (Miller et al. 1978; Mercer et al. 1991). This fact is of particular importance for the respiratory units (and the taper accounts for differences less than 1% in the radii).

The mass transport coefficient K_a, describing any uptake (loss) of a particular gas, depends on the mucus and tissue composition, such as mucus thickness, chemical composition of tissue etc.. The uptake of oxygen in the conducting airways can be set to 0, since in the acini it can be assumed to be 0.0057 mm^3 mm^{-2} s^{-1} (Lechner 1978). This is quite different for other types of gas. For a high reactive gas such as ozone, there is an uptake in the tubes of the bronchial tree, and from ozone dosimetry a value of 0.25 was determined. In the acini a mass transport coefficient of 1.36 is given (Miller 1978).

In conclusion, the following assumptions were made for this particular model:
1. Conducting airways do not change dimensions with breathing.
2. Three types of acini are differentiated.
3. No hysteresis of volume change of acini between inhalation and expiration is included.
4. Radial concentration differences are not considered in the airways as well as in the acini.
5. The interactions of convection, diffusion and uptake are approximated by an effective diffusion coefficient K_{aw}.

8.8
Implementation and Run-Times

All the necessary implementations were C-coded in a specific program (LUNGO2FUNCT). The program opens the structural data file LUNG.DAT and stores this information into memory. The user enters the degree of tidal volume and the

Table 1. Pressure/volume relationships of model

P (cmH$_2$O)	dV tot.	Vol. (ml) tot. ac.	Vol. (ml) mean ac.	Vol. (mm^3)	Ac.rad. (mm)
0	0	1.3	0.61	0.305	0.5
5	1.9	3.2	2.51	1.25	0.8
10	3.44	4.74	4.05	2.02	0.95
15	4.7	6.0	5.52	2.74	1.05
20	6.2	7.69	7.00	3.5	1.14
30	7.2	8.5	7.81	3.90	1.18

number of breathing cycles. At the beginning of the last cycle, when stable conditions are achieved, another file is opened for storage of results.

The equation above can be solved to determine the concentration in one segment at time step i+1 without the knowledge of the concentration at the next time step. Such an approach is strongly restricted in space and time. As a consequence, serious calculation errors such as fluctuations towards higher or lower concentration values may result during run time. Therefore, the main stem bronchi were divided up into sets of smaller sequential elements and the number of time steps was set to 3000, typically 10,000 iterations were used. For a higher accuracy of the model, the Crank-Nicholson approach was suggested, which permits more stability due to averaging the implicit and explicit mass balance (Press et al. 1988).

Running the program means that for 4000 segments and 2000 acini equations must be solved more than 10,000 times for inspiration and in addition for expiration. It takes up to 5–10 complete runs to have stable conditions in all segments. This is a task only useful to be run on very strong computers. Run time on a server (SGI Challenge, 150 MHz, R4400 Processor) takes about 2 s for one run; for 10,000 iterations this makes about 30 min; and for ten complete cycles it takes approximately 5 h.

Variations of this program have been developed to study flow, Reynold's numbers and resistance (LUNGFLOW) and ozone take-up (LUNGO3FUNCT).

9 Model Predictions

9.1
Convection and Reynolds Numbers

For the computer model of lung, the following Re numbers, using v and the width of the tubes, were calculated with the program LUNGFLOW. For normal breathing (model pressure 5–15 cmH$_2$O, 3 ml):
- At trachea d=3.40 mm, speed=1101 mm/s, Re=260
- At 1st left bronchius d=1.80 mm, speed=653 mm/s, Re=81
- At 1st right bronchius d=2.10 mm, speed=629 mm/s, Re=91
- At 2nd left bronchius d=2.10 mm, speed=379 mm/s, Re=55
- At 2nd right bronchius d=2.20 mm, speed=504 mm/s, Re=77
- At right intermediate lobe:
    - At stem of lobe d=1.8 mm, speed=120 mm/s, Re=22
    - At bronchial terminus d=0.3 mm, speed=1.6 mm/s, Re=0.034

Breathing at maximum vital capacity (model pressure of 0–30 cmH$_2$O, 7.2 ml):
- At trachea d=3.40 mm, speed=2644 mm/s, Re=625
- At 1st left bronchius d=1.80 mm, speed=1567 mm/s, Re=196
- At 1st right bronchius d=2.10 mm, speed=1511 mm/s, Re=221
- At 2nd left bronchius d=2.10 mm, speed=909 mm/s, Re=132
- At 2nd right bronchius d=2.20 mm, speed=1211 mm/s, Re=185
- At right intermediate lobe:
    - At stem of lobe d=1.8 mm, speed=288 mm/s, Re=54
    - At bronchial terminus d=0.3 mm, speed=4.0 mm/s, Re=0.08

From the above numbers it is obvious that a laminar flow can be expected throughout the model. The occurrence of turbulence (Re2100) is even for breathing at maximal vital capacity unlikely.

9.2
Oxygen and Ozone Mass Transport

Figure 45 gives the pressure required to maintain the flow predicted by the model due to the law of Poiseuille. A trajectory was defined from the trachea down to an terminating bronchiolus in the diaphragmatic lobe as indicated in Fig. 40. From order

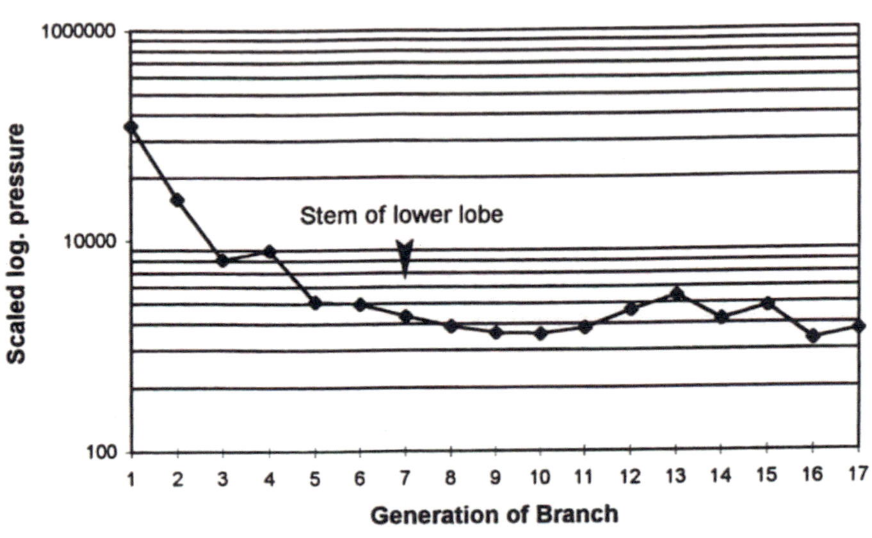

Fig. 45. Pressure required to maintain the flow predicted by the model. A trajectory was defined from the trachea down to an terminating bronchiolus in the diaphragmatic lobe as indicated in Fig. 40

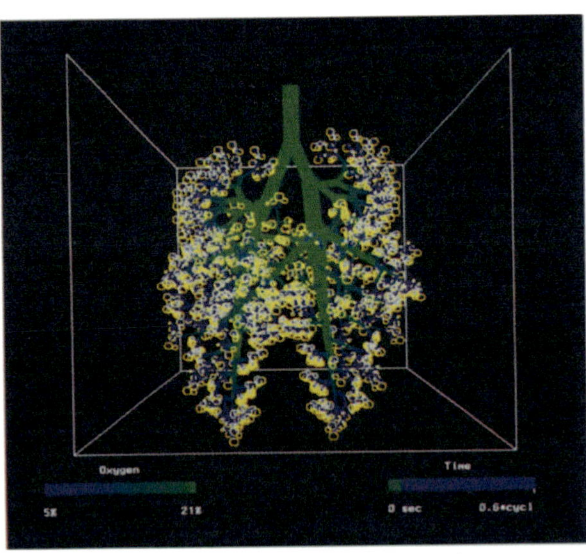

Fig. 46. Example from run-time of the functional computer model at the first breathing cycle. Gas concentration is indicated by coloring the segments and the increase of oxygen concentration of the main stem bronchi is visualized

86

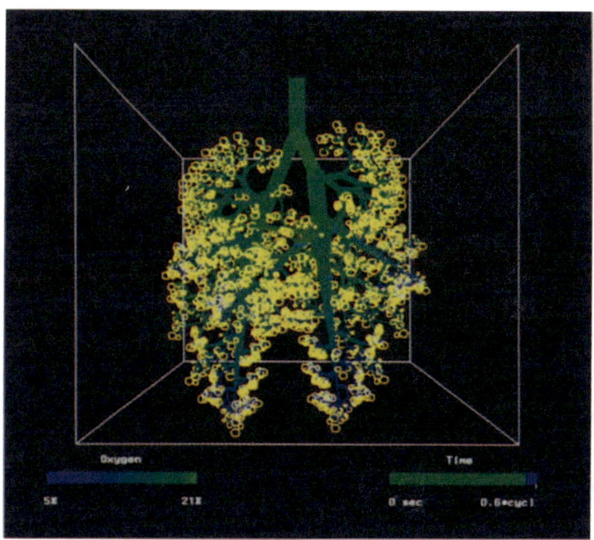

Fig. 47. Oxygen concentration at the end of expiration, the mean oxygen partial pressure is about 15% in the main stem bronchi due to oxygen takeup. Higher than average oxygen concentration in parts of the right lobe are a consequence of an insufficient number of iterations

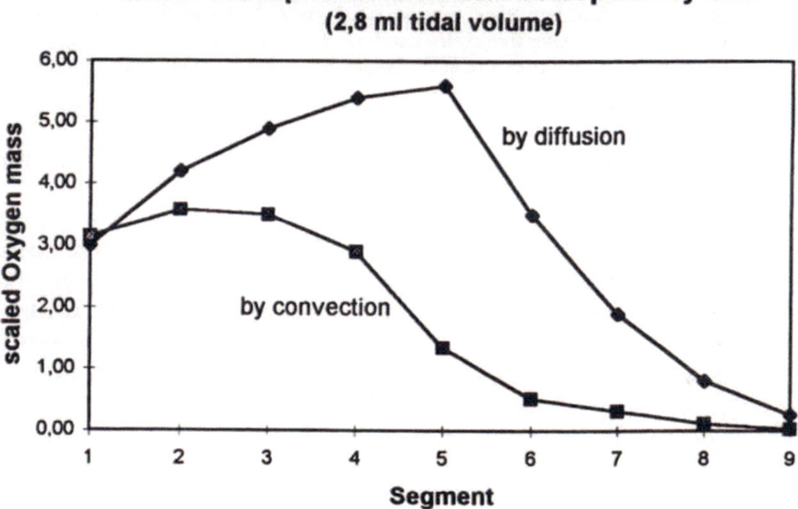

Fig. 48. Amount of oxygen mass transported through the segments of an acinus. Mass transport is about equal for convection and diffusion when entering the acinus, but the fast decrease in flow velocity drops the rate for convectional mass transport. Diffusion depends on the cross-sectional areas of the individual segments and increases towards the center of the acinus. Assymetry is due to oxygen takeup

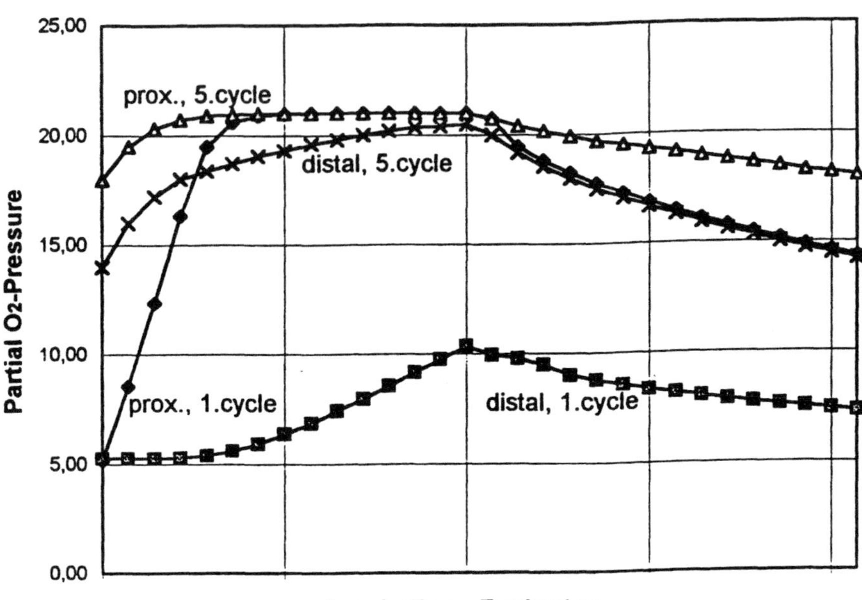

Fig. 49. Oxygen pressure at a proximal and distal located acinus, as indicated in Fig. 40. The distal acinus gets the same amount of flow, but "sees" a larger dead space. This can also be described by different transit times for both units. After five cycles a stable condition of the model is reached. This is indicated by equal amounts of partial pressure at the beginning and end of the cycle

5 on, the stem of the lobe, the pressure required remains despite some fluctuations constant.

Figures 46 and 47 give examples from run-time at the first breathing cycle. Oxygen enters the model as a bolus, since initially all values have been set to 5.26% partial pressure. Gas concentration is indicated by coloring the segments and by that the increase in oxygen concentration of the main stem bronchi is visualized. Figure 47 documents the situation at the end of expiration; the mean oxygen partial pressure is about 15% in the main stem bronchi due to oxygen take-up. There is still a gradient towards the distal segments, which is partly equalized if more than one cycle is calculated. The higher oxygen concentration in parts of the right lobe is a consequence of insufficient iterations (100 in this case).

Figure 48 documents the amount of oxygen mass transported through the segments of an acinus. This is a type 2 acinus, located near the stem of the intermediate lobe (see Fig. 40). Data have been taken at the end of the fifth inspiration cycle. Mass transport is about equal for convection and diffusion when entering the acinus, but the fast decrease in flow velocity quickly drops the rate of convectional mass transport. Diffusion, which, however, is dependent on the cross-sectional areas of the individual segments, increases towards the center of the acinus. Diffusional mass transport drops

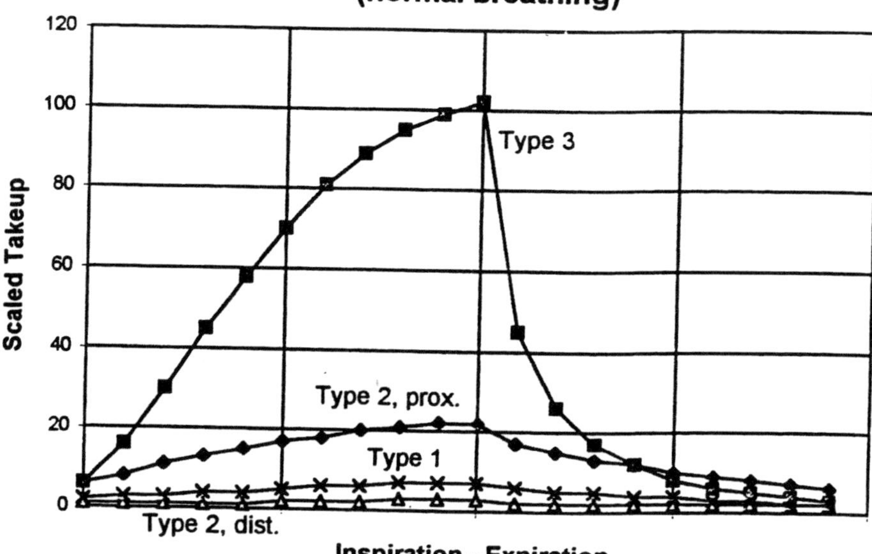

Fig. 50. Oxygen takeup for four individual respiratory units, three types located proximally and a type 2 also located distally. The exponential increase of takeup of larger acini, due to the large alveolar surface, is indicated. Maximum takeup occurs at the end of inspiration; assymetry is due to oxygen takeup

at a higher rate to the periphery of respiratory units. This is due to the absorption of oxygen, which also has an effect on the convectional mass transport.

Figure 49 represents the oxygen pressure at a proximal and distal located acinus, as indicated in Fig. 40. With the oxygen bolus entering the trachea, the proximal unit is quickly filled with oxygen, whilst the distal unit reaches its maximum of oxygen pressure at the very end of to the first inspiration cycle. The distal acinus gets the same amount of flow, but "sees" a larger dead space. This can be also described by the different transit times of the two units. After five cycles a stable condition of the model is reached and a typical pattern of partial pressures is achieved over a breathing cycle. This is indicated by equal amounts of partial pressure at the begin and end of the cycle.

Figure 50 compares the oxygen take-up for four individual respiratory units, three types located proximally and type 2 located also distally. Data have been taken from the fifth breathing cycle and 10,000 iterations performed for each cycle. Indicated is the exponential increase in take-up of larger acini, due to the large alveolar surface. Maximum take-up occurs at the end of inspiration; asymmetry is due to oxygen take-up. The location of acini also plays a significant role beside the size.

Figure 51 gives an interesting view of the discrete character of the model. At the beginning of the fifth breathing cycle the take-up of each acinus was constantly summed up. These quantities have been classified and scaled. Three discrete maxima are documented as a consequence of three types of acini modeled. Variance of these maxima is due to different locations within the bronchial tree.

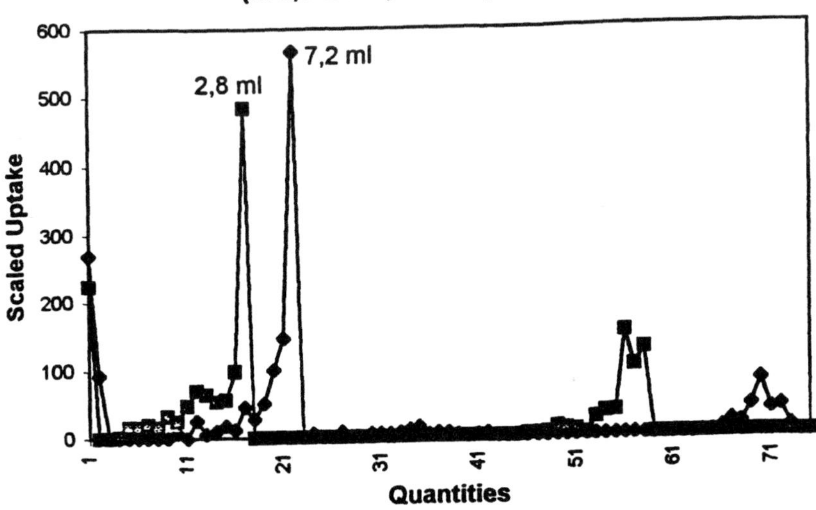

Fig. 51. Classified quantities of oxygen takeup. Three discrete maxima result form three types of acini modeled. Variance of these maxima is due to different locations within the bronchial tree

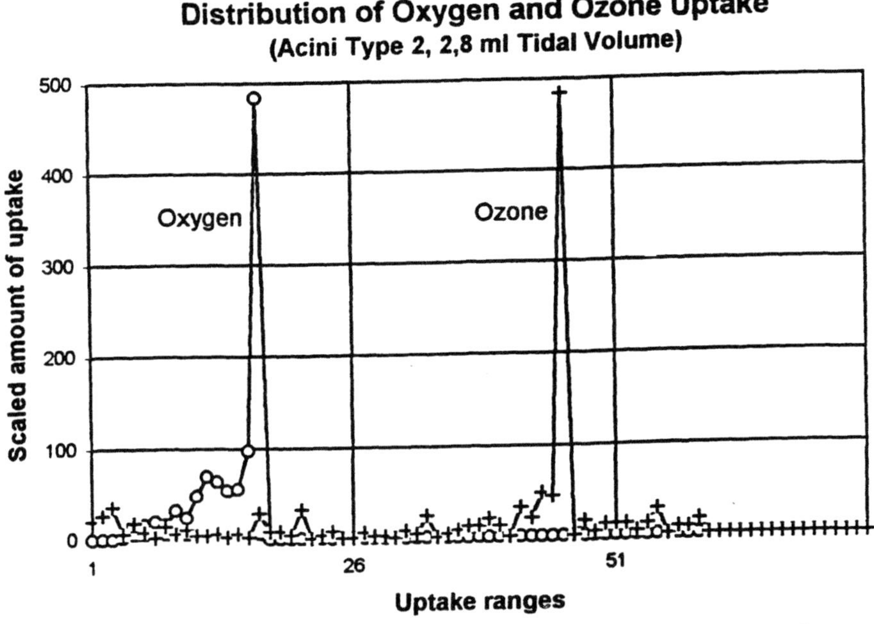

Fig. 52. Classified quantities of oxygen versus ozone takeup. The oxygen takeup, after reaching a maximum, drops to zero. This is in contrast to the ozone takeup, showing a much higher variance. The effect of variations in transit times are increased by ozone absorption in the bronchial tree

90

Figure 52 compares the uptake of oxygen versus ozone. Only respiratory units of type 2 have been used for statistics. The oxygen take-up, after reaching a maximum, drops to zero. This is in contrast to the ozone take-up, showing a much higher variance and smaller values which cover a wide range of amounts. This is a consequence of the locations of acini throughout the bronchial tree, since oxygen is already partly absorbed in the conductive part of lung, but oxygen is transported over longer paths without loss. By that, the variations in transit times are increased for ozone. Of interest is the absorption of higher amounts than average of ozone in some respiratory units, which are possible sites of tissue damage.

10 Discussion of Structural Modeling and Functional Simulation

10.1
Summary of Morphological Modeling

A computer model was developed that concerns the structure of the rat lung, which has about 4000 bronchial segments. Casts of a half-adult rat were studied which match the developmental stage of the histological material (see Chaps. 2–5). The method of investigation suggested here is to first measure the main branches of a lung exactly and analyze selected parts of the bronchial tree in detail and, secondly, to use obvious regularities and similarities present in the lung structure to complete a computer model by means of fractal graphics. Stochastic restriction by using measured data of the main branches and a matching against measured data improves the degree of proximity. As a result, a computer model of a complete bronchial tree is designed to study physiology and function after additional implementation of the necessary physical issues.

Modeling of the lungs branching pattern has been a subject of the mathematical treatment by fractal geometry, including famous workers such as Mandelbrot (1983). The essence of calculating such forms lies in the use of recursive algorithms. For the description of plant structure and development Lindenmayer (1983) introduced L-systems. The L-system is used to simulate the growth of structures by reproduction with the help of fractal templates (Rozenberg and Salomaa 1992; McCormick and Mulchandani 1994) and is used here for the modeling of self-similar parts of the bronchial tree.

The essential result of the analysis of casts was that the lung cannot be described with a simple fractal branching pattern. Instead, the individual lung lobes exhibit different dichotomous, symmetric and asymmetric-monopodial branchings. For the asymmetric forms, a transition in the bifurcation from asymmetry to dichotomy was observed and modeled. From this point of view it is also obvious that lung structures cannot be described by a simple fractal number.

10.2
How Could the Structure of the Bronchial Tree Be Explained?

Measurements of the lung airways demonstrate that the bronchial tree shares features common to fractal objects such as heterogeneity, self-similarity and lack of a characteristic scale. This has led to various ordering schemes, which in return makes it a problem to compare data on this subject.

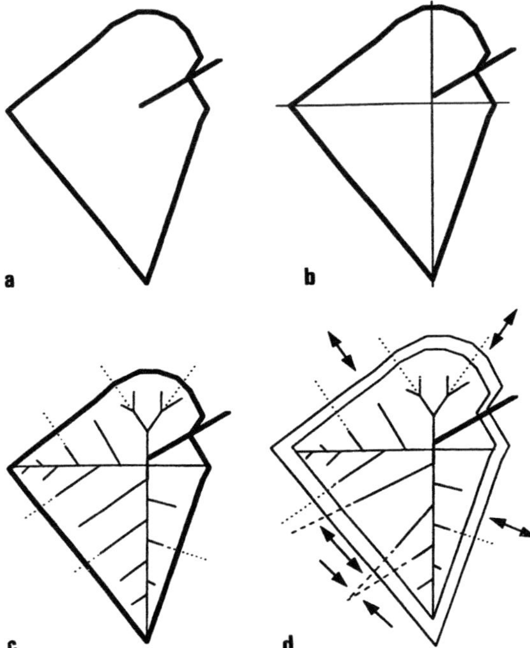

Fig. 53a–d. Construction scheme that suggests the morphogenesis of the bronchial tree. **a** Outer boundary of the left lung lobe. **b** Division of area by lines crossing the extremes of the outer boundary. **c** Constructions of bronchi oriented perpendicular to outer boundary. **d** Vectors indicating the functional consequences of airspace orientation for movement and force during breathing

From a theoretical standpoint, optimal airflow requires specific length-to-diameter relations and angular dimensions. Moreover, an optimized filling of the available lung volume and of all acini would be desirable. However, one must consider that during morphogenesis in in utero airway differentiation occurs without air or fluid exchange. It is therefore likely that other constraints play a role in the early stage of lung development. It was suggested that mechanical forces resulting from the development of adjacent structures and extrinsic intrauterine forces might be involved (Liggens 1984).

Despite the lack of knowledge about the amount of force acting onto the embryonic lung, it is reasonable to assume that the mechanical forces are equal in amount and are perpendicular to the surface of lung within the space available. Another reasonable assumption would be the functional need for an optimized filling of the available volume. Thus, the outer form of the lung must be an important constraint for the developing bronchial tree.

The outer boundary of a lung lobe of rat can be represented in a planar form such as in Fig. 53a. From here, the following receipt can be given to construct the basic geometry of the branching pattern of the lung lobe:

- Find x/y extrema (left/right, top bottom), link these extreme points by straight lines.
- Cut each line into segments of equal length, draw lines from each segment perpendicular to the outer boundary, and in alternating left/right orientation.

– Apply an dichotomous branching towards the boarder.

It is obvious that from this simple (geometric) construction schema the basic conductive pattern of the rat bronchial tree can be constructed (Fig. 53). This pattern explains as well the three types of branching patterns found:
– The dichotomous form in the apical part of lung is due to a round boundary. Forces act in parallel and over a wide angle onto all branches.
– The symmetric form of branching of the intermediate part of lung, ending in a tip of a regular triangle. Forces act equally and in parallel onto the branches.
– The asymmetric form of branching in the lower part of lung, ending in a tip of an asymmetric triangle. Again, forces act in parallel onto these branches.

The regular form of the lower lung lobe with equal branching angles of the main stems is hence explained as a consequence of the outer form.

During inspiration and expiration, the expansion and change of volume taking place in the respiratory units also generates inner and outer forces. Due to optimal space filling and bronchi being oriented in line, featuring an identical construction schema, these functional forces will act on the lung tissue such that they are equal in amount throughout the lung and that the force gradients and vectors keep their orientation. Otherwise there would be additional stress factors acting onto certain parts of the lung, causing tidal resistance or even ruptures (Fig. 53d). Another interesting fact in the construction scheme is that the conductive part of the bronchial tree can remain largely unchanged in form and in orientation during breathing.

10.3
Functional Predictions

Related to an optimized volume filling of the lung by the conductive part of the bronchial tree is the assumption that every terminating bronchiolus transports about equal quantities of gas into the acinus. This is in line with equal pressure gradients all over the lung. But, differences in gas concentration between the various respiratory units were predicted as a function of different transit times, since every acinus "sees" a different amount of dead space.

Depending on breathing frequency and amount of tidal volumes, typical constellations of partial pressure gradients occur. Computer visualization helps to elucidate these patterns. In principle, this holds for all other kind of gases involved, such as O_2, CO_2 or N_2. Different lengths of the bronchial pathways directly affect reactive gases such as ozone, which are partly absorbed during transit through the conductive part. The variance of gas up-take is increased. In other words, since oxygen is transported without major loss towards the respiratory units, the variance in up-take is mainly limited to differences in transit times.

To summarize these results, the following statements can be made as a consequence of the model predictions:
– No turbulence is present throughout the model at any tidal volume.
– Pressure differences to maintain the flow predicted would be constant from the stem of a lobe on.
– Every acinus gets the same amount of total ventilation.

- The effective ventilation with oxygen is, however, different due to different transit times. Every acinus "sees" a different amount of dead space.
- Typical gradients of gas concentration are generated for different tidal volumes
- Differences in effective ventilation are, in particular, obvious for ozone uptake by absorption in the bronchial tree.

10.4
Outlook

In addition to the use of confocal microscopy and image fusion for the investigation of the respiratory part of lung, the investigation of casts is perhaps still the most effective way to study the macroscopic, air-conducting branching pattern of smaller species.

For forms larger in space, such as the human lung, recent progress in HRCT has made 3-D imaging of the conductive part of the bronchial tree possible. The human lung would require an investigation of as many as 750,000 bronchial segments in the conductive part and multiprocessors or supercomputers would be desirable to design and work with such a model. Functional models used so far are based on a simplified, trumpet model only (e.g., Fleming et al. 1995). The necessary methodical steps described here may serve as a blueprint to develop such an appropriate model of the human lung, which would have great impact on medical applications.

11 Summary

Up to now it has been technically difficult to design a computer model of the complete bronchial tree of lung with all structural details. However, with the progress which has been made in 3-D image acquisition, processing, visualization, and computer modeling, such a representation of mammalian lungs has come within reach.

An initial set of hierarchical computer data representations has been envisioned here to design a model of the bronchial tree of a rat lung. Three levels of structural organization of the lung are taken into account: (1) the respiratory units (acini) at microscopic resolution, (2) the main stem macroscopic bronchi supplying for the lung lobes, and (3) the bronchial airway segments of the lung lobes. Each of these levels requires specific techniques for imaging, analysis, and modeling. The underlying idea of these investigation is to model a realistic template of the bronchial tree suitable for functional simulation.

Mechanical sectioning in conjunction with traditional microscopic techniques only delivers two-dimensional views of a structure which is essentially three-dimensional. Highly resolved, 3-D analysis of an acinus requires imaging of quite a large cube and a high resolution to resolve structural details. The imaging method developed here and described in Chap. 2 uses a computerized, large-area scanning technique applied to thick serial sections, which are resolved axially by optical tomography. This guarantees for an extended field of view, a high resolution laterally and axially, and a reduced workload for image alignment and, hence, less risk for generating disconnected structures.

In Chap. 3 the branching pattern of the acinar pathways is analyzed with a topology software. The branching pattern was presented in a 3-D view and plotted in a 2-D fashion. Measuring the diameter and areas of the ducts along the center lines gave a summed curve of cross sections and reveals a Gaussian distribution with a maximum at about half a diameter from the origin. Numerous investigations show that alveolar ducts are completely ensheathed with alveoli as well as alveolar sacs as the blind-ending terminations of alveolar ducts. This Gaussian distribution, contrary to common belief, indicates an asymmetric branching pattern within the acinus but not ducts of equal length and symmetric branching. In a way, asymmetry of the respiratory units resembles the asymmetry in the conductive part of the bronchial tree of rat and mouse, as studied later in this work. The knowledge of this distribution is the basis for modeling the functional issues of respiratory units. The aforementioned structural analysis is tightly coupled with computer visualistics as a tool to control and guide the individual processing steps and to give insight into the 3-D nature of this complex design, which is the subject of Chap. 4.

A computer model describing the structure of the conductive part of lung, which has about 4000 individual bronchial segments or airway tubes, is then developed. Casts of a half-adult rat were studied which match the developmental stage of the histological material. The method of investigation suggested in Chap. 6 is to first measure the main branches of a lung exactly and analyze selected parts of the bronchial tree in detail and, secondly, to use obvious regularities and similarities present in the lung structure to complete a computer model by means of a fractal graphics, which is the main topic of Chap. 7. By stochastically restricting the data measured from the main branches and matching them against the measured data improves the degree of proximity.

The essential result of the analysis of casts was that the lung cannot be described with a simple fractal branching pattern. Instead, the individual lung lobes exhibit different dichotomic, symmetric and asymmetric-monopodial branchings. For the asymmetric forms, a transition in the bifurcation from asymmetry to dichotomy was observed and modeled. From this point of view it is also obvious that lung structures cannot be described by a simple fractal number.

In contrast to the belief that the design of the conductive part of lung of smaller mammals can be described with a trumpet model, the findings reported here document a strongly monopodial branching pattern with the functional consequence of a variation of dead space between the trachea and the acini. This nondichotomic structural design finds a continuation within the respiratory units, as the necessary requirement for an optimal space filling and dense packing which cannot be achieved by a dichotomic branching only.

In Chap. 8 the computer model of a complete bronchial tree is then used to study physiology and function after additional implementation of the necessary physical issues. All bronchial segments are handled as individual objects in a finite-element model. Gas transport is described by convection, diffusion and uptake. The corresponding differential equations are solved iteratively. Computer visualization helps to elucidate concentration patterns.

We can assume that the morphological and functional properties of all respiratory units are about equal, designed for an efficient up-take at maximum oxygen concentration. However, depending on the location of acini, from proximal to more distal and on the breathing volume, these are filled with oxygen to different extents. In other words, every acinus gets the same amount of total ventilation, but the effective ventilation with oxygen is different due to different transit times. As a result, specific patterns of oxygen concentration are brought about within the bronchial tree, as demonstarted in Chap. 9.

The different ventilation of the acini with oxygen gives rise to a possible explanation of the well-known difference between the morphological diffusion capacity of lung and the physiological diffusion capacity. This explains the paradox of Weibel.

There is no doubt that from the study ofrat lung, finding ways of generating the necessary image data, constructing a hierarchical structural model, and implementing computational physics, the computer model presented here can serve not only as a blueprint for a model of the human lung, but may also initiate similar studies of other organs.

98

12 References

Agard D, Hiraoka Y, Shaw P, Sedat JW (1989) Fluorescence microscopy in three dimensions. Methods in cell biology, vol 30. Academic, pp 353–377

Art J (1990) Photon detectors for confocal microscopy. In: Pawley J (ed) Handbook of biological confocal microscopy. Plenum, New York, pp 127–139

Alonso M, Finn EJ (1967) University physics, vol 2. Addison-Wesley, Reading

Aschoff L (1935) Über den Lungenacinus. Frankfurt Z Pathol 48:449–455

Baumann MA, Schwebel T, Kriete A (1993) Dental anatomy portrayed with microscopic volume investigations. Computerized medical imaging and graphics, vol 17. Pergamonn, Oxford, pp 221–228

Bhawalkar JD, He GS, Prasad PN (1995) Three-photon induced upconverted fluorescence from an organic compound: application to optical power limiting. Optics Commun 119:587–590

Bloch P, Udupa UK (1983) Application of computerized tomography to radiation and surgical planning. Proc IEEE 71:351–355

Boyden EA (1971) The structure of the pulmonary acinus in a child of six years and eight months. Am J Anat 132:275–300

Boyden EA (1980) The mode of origin of the pulmonary acini and respiratory bronchioles in the fetal lung. Am J Anat 141:317–328

Brakenhoff GJ, Blom P, Barends P (1979) Confocal scanning light microscopy with high aperture immersion lens. J Microsc 117:219–232

Brakenhoff GJ, Van der Voort HTM, Van Spronsen EA, Nanninga N (1989) Three-dimensional imaging in fluorescence by confocal scanning microscopy. J Microscopy 153:151–159

Brakenhoff GJ, Visscher K (1990) Bilateral scanning and array detectors. J Microsc 165:139–146

Bron C, Gremillet P (1992) 3-D reconstructions by image processing of serial sections in electron microscopy. In: Kriete A (ed) Visualization in biomedical microscopies – 3-D imaging and computer applications. VCH, Weinheim, pp 75–105

Cabral B, Cam N, Foran J (1995) Accelerated volume rendering and tomographic reconstruction using texture mapping hardware. Internal Report Silicon Graphics, Mountain View, CA

Carlsson K, Lunddahl P (1991) Three-dimensional specimen recording and interactive display using confocal laser microscopy and digital image processsing. Machine Vision Appl 4:215–225

Chang C (1985) General concepts of molecular diffusion. In: Engel LA, Paiva M (eds) Gas mixing and distribution in the lung. Dekker, New York, pp 1–22

Chen H, Sedat JW, Adard JA (1990) Manipulation, display and analysis of three- dimensional biological images. In: Pawley J (ed) Handbook of biological confocal microscopy. Plenum New York, pp 141–150

Chen LS, Herman GT, Reynolds RA, Udupa JK (1985) Surface shading in the cuberille environment. IEEE Comput Graph Appl 5:33–43

Cheng PC, Kriete A (1995) Image contrast in confocal light microscopy. In: Pawley J (ed) Handbook of biological confocal microscopy. Plenum, New York, pp 281–308

Cheng PC, Achary R, Lin TH, Samarabandu JK, Wang G, Shinozaki DM, Berezney R, Meng C, Tarng WH, Liou WS, Tan TC, Summers RG, Kuang H, Musial C (1992) 3-D image analysis and visualization in light microscopy and X-ray micro-tomography. In: Kriete A (ed) Visualization in biomedical microscopies – 3-D imaging and computer applications. VCH, Weinheim, pp 361–398

Cheng PC, Lin TH, Wu, WL, Wu JL (eds) (1994) Multidimensional microscopy. Springer, Berlin Heidelberg New York

Conan V, Howard V, Jeulin D, Renard D, Cummins P (1990) Improvement of 3D confocal images by geostatistical filters. In: Elder HY (ed) Transactions of the Royal Microscopy Society, vol 1. Hilger, Bristol, pp 281–284

Conchello JA, Hanssen EW (1990) Enhanced 3-D reconstruction from confocal scanning microscope images. Deterministic and maximum likelihood reconstructions. Appl Optics 29(26):3795–3804

Cookson MJ, Davies CJ, Entwistle A, Whimster WF (1993) The microanatomy of the alveolar duct of the human lung imaged by confocal microscopy and visualized with computer-based 3D reconstruction. Comput Med Imaging Graph 17(3):201–210

Cogswell CJ (1993) High resolution confocal microscopy of phase and amplitude objects. In: Cheng PC, Lin TH, Wu WL, Wu JL (eds) Multidimensional microscopy. Springer, Berlin Heidelberg New York, pp 87–102

Cogswell CJ, Sheppard CJR (1992) Conventional and confocal DIC-imaging. J Microsc 165:81–101

Cox G (1993) Photons under the microscope: Marvin Minsky – the forgotten pioneer. Aust Electron Microsc Newsletter 38:4–10

Cox IJ, Sheppard CJR, Wilson T (1982) Superresolution in confocal fluorescent microscopy. Optik 60:391–396

Debrin RA, Carpenter L, Hanrahan P (1988) Volume rendering. Comput Graph 22:65–74

Deitch JS Smith KL, Swann JW, Turner JN (1990) Parameters affecting imaging of HRP reaction product in the confocal scanning laser microscope. J Microsc 160:265–278

Dubois AB (1964) Resistance to breathing. In: Fenn O, Rahn H (eds) Respiratin, vol 1. American Physiological Society, Washington,p 451–461 (Handbook of Physiology)

Duke PJ, Michette AG (eds) (1990) Modern microscopies – techniques and applications. Plenum, New York

Dupuy O (1986) La perception visuelle. Vis Res 8:1507

Egger MD, Petran M (1987) New reflected light microscope for viewing unstained brain and ganglian cells Science 157:305–307

Emmet A (1992) Down to earth – practical application of virtual reality for commercial use. Comput Graph World (March):46–54

Engel LA, Paiva M (eds) (1985) Gas mixing and distribution in the lung. Dekker, New York

Fenn WO, Rahn H (eds) (1964) Handbook of physiology, vols 1, 2. American Physiological Society, Washington

Fleming JS, Nassim M, Hashish AH, Bailey AG, Conway J, Holgate S, Halson P, Moore E, Martonen TB (1995) Description of pulmonary deposition of radiolabeled aerosol by airway generation using a conceptual three dimensional model of lung morphology. J Aerosol Med 8(4):341–356

Forsgren PO (1990) Visualization and coding in three-dimensional image processing. J Microsc 159(2):195–202

Frank J (1980) The role of correlation techniques in computer image processing. In: Hawkes PW (ed) Computer processing of electron microscope images. Topics in current physics. Springer, Berlin Heidelberg New York, pp 187–222

Franke G (1965) Bildgütekriterien. Optik 23(1):20–218

Frieder G, Gordon D, Reynolds RA (1985) Back-to-front display of voxel-based objects. IEEE Comput Graph Appl 5(1):52–60

Frieser H (1975) Photographic image recording – fotografische Informationsaufzeichnung. Focal-Oldenbourgh, Munich

Fritjers D, Lindenmayer A (1974) A model for growth and flowering of Aster novaeangliae on the basis of table (1,0) L-systems. In: Rozenberg G, Salomaa A (eds) L-systems, lecture notes in computer science, vol 15, Springer, Berlin Heidelberg New York, pp 24–52

Gerald CF, Weatley PO (1970) Applied numerical analysis, 5th edn. Addison-Wesley, Reading

Goldstein RA, Nagel R (1971) 3-D visual simulation. Simulation 16(1):25–31

Gordan D, Reynolds RA (1985) Image space shading of 3-D objects. Comput Vision Graph Image Proc 29:361–376

Gouroud H (1971) Continuous shading of curved surfaces. IEEE Trans Comput 20(6):623–629

Gratton E, Van deVeen MJ (1990) Laser sources for confocal microscopy. In: Pawley J (ed) Handbook of biological confocal microscopy. Plenum, New York, pp 69–97

Hansen FG, Ampaya EP (1986) Lung morphometry: a fallacy in the use of the counting principle J Appl Physiol 37:951–954

Hansen JE. Ampaya EP (1975) Human air space shapes, sizes, areas and volumes. J Appl Physiol 38:990–995

Haugland RP (1989) Molecular probes: handbook of fluorescent probes and research chemicals. Molecular Probes, Eugene

Heafeli-Bleuer B, Weibel ER (1988) Morphometry of the human pulmonary acinus. Anat Rec 220:401–414

Hell S, Stelzer EHK (1992) Properties of a 4Pi confocal fluorescence microscope. J Opt Soc Am 9:2159–2166

Hell S, Lethonen E, Stelzer EHK (1992) Confocal fluorescence microscopy: wave optics considerations and applications to cell biology. In: Kriete A (ed) Visualization in biomedical microscopies – 3-D imaging and computer applications. VCH, Weinheim, pp 145–160

Hell SW, Bahlmann K, Schrader M, Soini A, Malak H, Gryczynski I, Lakowicz JR (1996) Three-photon excitation in fluorescence microscopy. J Biomed Optics 1(1):71–74

Hellmuth T, Seidel P, Siegel A (1988) Spherical aberration in confocal microscopy. Proc SPIE 1082:28–32

Herman GT (1981) Three-dimensional imaging from tomograms. In: Höhne KH (ed) Digital image processing in medicine. Lecture notes in medical informatics, vol 15. Springer, Berlin Heidelberg New York, p 93

Herman GT, Liu HK (1979) Three-dimensional display of human organs from computed tomograms. Comp Graphics Image Proc 9:1–21

Herman GT, Udupa JK (1983) Display of 3-D digital images: computational foundations and medical applications. IEEE Comput Graph Appl 3:39–46

Herrmann KH, Krahl D (1982) The DQE of electronic image recording systems. J Microsc 127:17

Hersh JS (1990) A survey of modeling representations and their application to biomedical visualization and simulation. VBC Conference 1990. IEEE Computer Society Press, pp 432–441

Höhne KH, Riemer M, Tiede U (1987) Viewing operations for 3-D tomographic grey level data. In: Computer assisted radiology (CAR 1987). Springer, Berlin Heidelberg New York, pp 599–609

Hodges LF (1991) Basic principles of stereographic software development, SPIE vol 1457 II, pp 9–17

Horsfield K (1985) Anatomical factors influencing gas mixing and distribution. In: Engel LA, Paiva M (eds) Gas mixing and distribution in the lung. Dekker, New York, pp 23–61

Horsfield K, Cumming G (1968) Functional consequences of airway morphology. J Appl Physiol 24(3):384–390

Huijsmanns DP, Lamers WH, Los JA, Strackee J (1986) Toward computerized morphometric facilities: a review of 58 software packages for computer-aided 3-D reconstruction, quantification, and picture generation from parallel serial sections. Anat Rec 216:449–470

Hyatt RE, Wilcox RE (1963) The pressure-flow relationship of the intrathoracic airways in man. J Clin Invest 42:1455–1465

Inoue S (1990) Foundations of confocal scanning imaging in confocal microscopy. In: Pawley J (ed) Handbook of biological confocal microscopy. Plenum, New York, pp 1–14

Katz L, Levinthal C (1972) Interactive computer graphics and representation of complex biological structures. Annu Rev Biophys Bioeng 1:465–504

Kaye BH (1989) A random walk through fractal dimensions. VCH, Weinheim

Kitaoka H, Takahashi T (1993) Relationship between the branching pattern of airways and the spatial arrangement of pulmonary acini – a re-examination from a fractal point of view. In: Nonnemacher TF, Losa GA, Weibel ER (eds) Fractals in biology and medicine. Birkhäuser, Basel, pp 116–131

Kriete A (1990) 4-D data acquisition and visualization methods in computer assisted microscopy. In: Elder HY (ed) Transactions of the Royal Medical Society. Hilger, Bristol, pp 323–326

Kriete A (ed) (1992) Visualization in biomedical microscopies – 3-D imaging and computer applications. VCH, Weinheim

Kriete A (1994) Image quality considerations in 2-D and 3-D microscopy li: Cheng PC, Lin TH, Wu Wl, Wu Jl eds.) Multidimensional Microscop. Springer NY

Kriete A, Aus HM (1986) On-line processing of transmission electron microscopic images. Pattern Recognition Lett 4 (366):285–292

Kriete A, Magdowski G (1990) Computerized 3-D reconstructions of serial sections in electron microscopy. Ultramicroscopy 32:48–54

Kriete A, Masters B (1990) Three-dimensional visualization of of the living cornea. In: Elder HY (ed) Transactions of the Royal Medical Society. Hilger, Bristol, pp 323–326

Kriete A, Pepping T (1992) Volumetric data representations in microscopy: Applications to confocal and NMR-microimaging In Kriete A (ed) Visualization in biomedical microscopies – 3-D imaging and computer applications. VCH, Weinheim, pp 329–357

Kriete A, Wagner HJ (1993) A new method for 4-D data representation in microscopy: application to neuroanatomical plasticity. J Microsc 169:27–31

Kriete A, Schwebel T (1996) 3D-TOP, a software package for the topological analysis of image sequences. J Struct Biol 116:150–154

Kriete A, Rohrbach S, Schwebel T (1992) Data representation and visualization in 4-D microscopy. In: Robb R (ed) Visualization in biomedical computing. SPIE vol 1808,. Chapel Hill, NC, pp 396–409

Kriete, A, Schwebel T, Duncker HR, Marko M (1996) Multiscale imaging, analysis and modeling of the pulmonary bronchial tree. J Zool (in press)

Lamers WH, Laan AC, Huijsmanns DP, Smith J, Los JA, Strackee J (1989) Deformation-corrected computer-aided 3-D reconstruction of immunohistochemically stained sections of embryonic organs. Eur J Cell Biol [Suppl 25] 48:103–106

Lansing D, Taylor L, Wang YL (eds) (1989) Methods in cell biology, vol 30, Academic, New York

Lechner AJ (1978) The scaling of maximal oxygen consumption and pulmonary dimensions in small mammals. Respir Physiol 34:29–44

Levoy M (1988) Display of surfaces from volume data. IEEE Comput Graph Appl 8(3):29–37

Levoy M (1990) A hybrid ray-trace for rendering polygon and volume data. IEEE Comput Graph Appl 3:33–40

Levoy M, Fuchs H, Pizer MS, Rosenmann J, Chaney EL, Sherouse GW, Interrante V, Kiel J (1990) Volume rendering in radiation treatment planning. VBC 1990, Atlanta. Proceedings IEEE. IEEE Computer Society Press, pp 4–10

Ley K, Pries AR, Gaethgens P (1986) Topological structure of rat mesenteric microvessel networks. Microvasc Res 32:315–332

Liggens GC (1984) Growth of the fetal lung. J Develop Physiol 6:237–248

Lindenmayer A (1968) Mathematical models for cellular interaction in development. J Theor Biol 18:280–315

Ling GN (1992) Can we see living structure in a cell? Scanning Microsc 6(2):405–450

Lobregt S. Verbeek PW, Groen FCA (1980) Three-dimensional sceletonization – principle and algorithm. IEEE Trans PAMI 2:75–77

Mandelbrot BB (1967) How long is the coast of Britain, statistical self similarity and fractional dimension. Science 155:636–638

Mandelbrot BB (1977) Fractals: form, chance and dimension. Freeman, San Francisco

Marek W, Ulmer WT (1995) Abhängigkeit des Strömungswiderstandes und des intrathorakalen Gasvolumens von der Strömung und der Atemfrequenz bei gesunden Erwachsenen. Pneumologie 49:410–412

Marko M, Leith A (1992) Contour-based surface reconstruction using stereoscopic contouring and digitized images. In: Kriete A (ed) Visualization in biomedical microscopies – 3-D imaging and computer applications. VCH, Weinheim, pp 45–73

Masters B (1992) Confocal ocular microscopy – a new paradigm for ocular visualization. In: Kriete A (ed) Visualization in biomedical microscopies – 3-D imaging and computer applications. VCH, Weinheim, pp 183–203

Masters B, Kriete A, Kukulies J (1993) Ultraviolett confocal fluorescence microscopy of the in-vivo cornea: redox metabolic imaging. Appl Opt 32(4):592–596

McCormick BH, Mulchandani (1994) L-system modeling of neurons. In: Robb R (ed) Visualization in biomedical computing. SPIE vol 2359, pp 693–705

Mercer RR, Laco JM, Crapo JD (1987) Three-dimensional reconstruction of alveoli in the rat lung for pressure-volume relationships. J Appl Physiol 62(4):1480–1487

Mercer RR, Anjilvel S, Miller FJ, Crapo JD (1991) Inhomogenity of ventilatory unit volume and its effects on reactive gas uptake. J Appl Physiol 70(5):2193–2205

Meyer F (1992) Mathematical morphology: 2-D to 3- D. J Microsc 165:5–28

Miller FJ, Menzel B, Coffin DL (1985) Similarity between man and laboratory animals in regional pulmonary deposition of ozone. Environ Res 17:84–101

Miller WS (1937) The lung. Thomas, Springfield

Montag M, Spring H, Trendelenburg MF, Kriete A (1990) Methodical aspects of 3-D reconstruction of chromatin architecture in mouse trophoblast giant nuclei. J Microsc 158:225–233

Moss VA (1992) Fundamentals of 3-D reconstruction of serial sections on a microcomputer. In: Kriete A (ed) Visualization in biomedical microscopies – 3-D imaging and computer applications. VCH, Weinheim, pp 19–44

Mossberg K, Arvidsson U, Ulfhake B (1990) Computerized quantification of immunoflourescence labelled axon terminals and analysis of co-localization of neurochemicals in axon terminals with a confocal scanning laser microscope. J Histochem Cytochem 38:179–190

Nelson TR, Manchester DK (1988) Modeling of lung morphogenesis using fractal geometries. IEEE Trans Med Imaging 7(4):321–327

Newman SA, Comper WD (1990) Generic physical mechanisms of morphogenesis and pattern formation. Development 110:1–18

Nonnenmacher TF, Losa GA, Weibel ER (eds) (1993) Fractals in Biology and Medicine. Birkhäuser, Basel

Oldmixon EH, Carlsson K (1993) Methods for large data volumes from confocal scanning laser microscopy of lung. J Microscopy 170(3):221–228

Oliver D, Hovis D (1994) Fractal Graphics. SAMS Publishing, Indiana

Otis AB, McKerrow CB, Bartlett RA, Mead J, McIlroy MB, Selverstone JJ, Radford EP (1956) Mechanical factors in the distrbution of pulmonary ventilation. J Appl Physiol, 8:427–443

Paiva M (1958) Theoretical studies of gas mixing in the lung. In: Engel LA, Paiva M (eds) Gas mixing and distribution in the lung. Dekker, New York, pp 221–285

Pawley J (ed) (1990) Handbook of biological confocal microscopy. Plenum, New York

Pawley J (ed) (1990) Handbook of biological confocal microscopy. Plenum, New York

Petran M, Hadravsky M, Egger MD, Galambos R (1968) Tandem-scanning reflected light microscope. J Opt Soc Am 58:661–664

Phalen RF, Yeh HC, Schum GM, Raabe OG (1978) Application of an idealized model to morphometry of the mammalien tracheobronchial tree. Anat Rec 190:167–176

Phong BT (1975) Illumination for computer generated pictures. Commun Az Cz Mz 18(6):311–317

Picaud S, Peichl L, Franceschini N (1990) Dye-induced photo-generation and photo-permeabilization of mammalien neurons in vivo. Vis Res 531:117–126

Press WH, Flannery BP, Teukolsky SA, Vetterling WT (1988) Numerical recipes. Cambridge University Press, Cambridge

Prisinkiewicz P, Lindenmayr A (1990) The algorithm beauty of plants. Springer, Berlin Heidelberg New York

Raya PS, Udupa JK, Barett WA (1990) A PC-based 3D imaging system: algorithms, software, and hardware considerations. Comput Med Imaging Graph 14(5):353–370

Renz W, Westermann R, Krüger W (1994) Interactive visualization of three-dimensional automata. Comput Phys 8(5):550–555

Rigaut JP Caravajal-Gonzalez S, Vassy J (1992) 3-D image cytometry In: Kriete A (ed) Visualization in biomedical microscopies – 3-D imaging and computer applications. VCH, Weinheim, pp 205–237

Rindfleisch E (1878) Lehrbuch der pathologischen Gewebelehre, 5th edn. Engelmann, Leipzig

Robb RA (1995) Three-dimensional biomedical imaging: principles and practice. VCH, Weinheim

Robb RA, Barillot C (1989) Interactive display and analysis of 3-D medical images. IEEE Trans Med Imaging 8:217–226

Roth SD (1982) Ray-casting for solid modeling. Comput Graph Image Processing 18:109–144

Rodriguez M, Bur S, Favre A, Weibel ER (1987) Pulmonary acinus: geometry and morphometry of the peripheral airway system in rat and rabbit. Am J Anat 180:143–155

Romeis B (1989) Mikroskopische Technik. Oldenbourgh, Munich

Ross BB (1956) Influence of bronchial tree on ventilation in the dog's lung as inferred from measurements of plastic cast. J Appl Physiol 30:349–367

Ross BB (1957) Influence of bronchial tree structure on ventilation in the dog's lung as inferred from measurements of a plastic cast. J Appl Physiol 10:1–14

Roth SD (1982) Ray-casting for solid modeling. Comput Graph Image Processing18:109–144
Rozenberg G, Salomaa A (eds) (1992) Lindenmayer systems. Springer, Berlin Heidelberg New York
Rubin SM, Whitted T (1980) A three-dimensional representation for fast rendering of complex scenes. Comput Graph 14:110–116
Russ JC (1988) Computer assisted microscopy. North Carolina State University Publication, Rawleigh
Sandison DR, Williams RM, Wells KS, Strickler J, Webb WW (1995) Quantitative fluorescence confocal laser scanning microscopy (CLSM). In: Pawley J (ed) Handbook of biological confocal microscopy. Plenum, New York, pp 39–54
Scherer PW, Shendalman LH, Greene NM, Bouhuys A (1975) Measurement of axial diffusivities in a model of the bronchial airways. J Appl Physiol 38(4):719–723
Schubert R, Höhne KH, Pommert A, Riemer M, Schiemann T, Tiede U, Lierse W (1993) Spatial knowledge representation for visualization of human anatomy and function. In: Barret HH, Gmitro AF (eds) Information processing in medical imaging. Proceedings of the IPMI 1993. Lecture notes in computer science, vol 687. Springer, Berlin Heidelberg New York, pp 168–181
Sernetz M, Justen M, Jestczemski F (1995) Dispersive fractal characterization of kidney ateries by three-dimensional mass-radius-analysis. Fractals 3(4):879–891
Sheppard CJR, Choudhury A (1977) Image formation in the scanning microscope. Opt Acta 24:1051–1073
Sheppard CJR, Wilson T (1978) Depth of field in the scanning microscope. Opt Lett 3:115–117
Shotton DM (1989) Confocal scanning optical microscopy and its applications for biological specimens. J Cell Sci 94:175–206
Shotton D (ed) (1993) Electronic light microscopy. Wiley-Liss, New York
Slutzky AS, Kamm RD, Drazen JM (1985) Alveolar ventilation at high frequencies using tidal volumes smaller than the anatomical dead space. In: Engel LA, Paiva, M (eds) Gas mixing and distribution in the lung. Dekker, New York, pp 137–176
Smith J, Jongenlen J, Lamers WH, Los JA, Strackee J (1989) Octree representation for 3-D image analysis. Eur J Cell Biol [Suppl 25]48:97–98
Stevens JK, Trogadis J, Mills L, Leitao C (1990) Three-dimensional volume investigation of serial confocal data sets. In: Elder HY (ed) Transactions of the Royal Medical Society. Hilger, Bristol, pp 375–380
Taylor DL, Nederlof M, Lanni F, Waggoner AS (1992) The new vision of light microscopy. Am Sci 80:322–335
Terasaki M, Dailey ME (1995) Confocal microscopy of living cells. In: Pawley J (ed) Handbook of biological confocal microscopy. Plenum, New York, pp 327–344
Tsien RY, Bacskai BJ (1995) Video-rate confocal microscopy. In: Pawley J (ed) Handbook of biological confocal microscopy. Plenum, New York, pp 459–478
Tsien RY, Waggoner A (1989) Fluorophores for confocal microscopy: photophysics and photochemistry. In: Pawley J (ed) Handbook of biological confocal microscopy. Plenum, New York, pp 169–178
Tuy HK, Tuy LT (1984) Direct 2-D display of 3-D objects. IEEE Comput Graph Appl 4:29–33
Udupa JK, Hung HM (1990) Surface versus volume rendering: a comperative assessment. VBC 1990, Atlanta. In:First conference on visualization in biomedical computing. Atlanta, IEEE Computer Society Press,Los Alamitos, pp 83–91
Ultman JS (1983) Gas transport in the conducting airways In: Engel LA, Paiva M (eds) Gas mixing and distribution in the lung. Dekker, New York, pp 63–136
Ultman JS, Blatman HS (1977) Logitudinal mixing in pulmonary airways. Analysis of inert gas dispersion in symetric tube network models. Respir Physiol 30:349–367
Valerius KP (1992) Bauprinzipien der Säugerlunge: Größenabhängige Veränderungen des konduktiven und respiratorischen Bronchialbaums bei mäuseverwandten Nagetieren. Thesis, University of Giessen
Van der Pol AN (1989) Neuronal imaging with collodial gold. J Microsc 155:27–59
Villringer A, Dirnagl U, Einhäupl KM (1992) Microscopical visualization of the brain in-vivo. In: Kriete A (ed) Visualization in biomedical microscopies – 3-D imaging and computer applications. VCH, Weinheim, pp 161–181
Visser T, Brakenhoff GJ (1991) The point response in fluorescence confocal microscopy. Optik 87:39–40

104

Voort, van der HTM, Brakenhoff GJ, Baarslag MW (1989) Three-dimensional visualization methods. J Microscopy 153(2):123–132

Voort, van der HTM, Brakenhoff GJ (1990) 3-D image formation in high aperture fluorescence confocal microscopy: a numerical analysis. J Microscopy 158(1):43–54

Wallen P, Carlsson KMossberg K (1992) CLSM as a tool for studying the 3-D morphology of nerve cells. In Kriete A (ed) Visualization in biomedical microscopies – 3-D imaging and computer applications. VCH, Weinheim, pp 110–143

Wang G, Liou W, Lin T, Cheng PC (1994) Image restoration in light microscopy. In: Cheng PC, Lin TH, Wu, WL, Wu JL (eds) Multidimensional microscopy. Springer, Berlin Heidelberg New York, pp 191–208

Ware R, LoPresti V (1975) Three-dimensional reconstruction from serial sections. Comput Graph 2:325–440

Webb WW (1990) Two photon excitation in laser scanning fluorescence microscopy. In: Elder HY (ed) Transactions of the Royal Microscopy Society. Hilger, Bristol, pp 445–450

Weibel ER (1963) Morphometry of the human lung. Springer, Berlin Heidelberg New York

Weibel ER (1983) Sinnvolle morphometrische Deskriptoren am Beispiel des respiratorischen Systems. Verh Anat Ges 77:165–170

Weibel ER (1993) Design of biological organisms and fractal geometry. In: Nonnemacher TF, Losa GA, Weibel ER (eds) Fractals in biology and medicine. Birkhäuser, Basel, pp 68–85

Weibel ER, Taylor CR (1988) Design of the mammalian respiratory system. Respir Physiol 44:1–164

Wells KS, Sandison DR, Strickler J, Webb WW (1989) Quantitative fluorescence imaging with laser scanning confocal microscopy. In: Pawley J (ed) Handbook of biological confocal microscopy. Plenum, New York, pp 39–54

Westermann R (1994) A multiresolution framework for volume rendering. Conference proceedings ICM, workshop on volume visualization. IEEE, pp 51–58

Wilson T (1989) Optical sectioning in confocal fluorescence microscopy. J Microsc 154:143–156

Wilson T, Carlini AR (1987) Depth discrimination criteria in confocal optical systems. Optik 76(4):164–166

Yeh HC, Schum GM, Duggan MT (1979) Anatomic models of the tracheobronchial and pulmonary regions of the rat. Anat Rec 195:483–492

Subject Index

Made in the USA
Monee, IL
07 July 2026

56552012R00074